"Jonathan Fisk has given us an excellent analysis of state and local fracking politics and policy-making that is informative, well written, and eminently readable. Especially useful is a concluding chapter that offers a number of policy prescriptions designed to facilitate state and local efforts to balance energy and environmental protection while recognizing the need for both transparency and public involvement."

—***Charles Davis***, *Colorado State University, USA*

"Fracking represents a test of the capacity of states and localities to be good environmental stewards given the very modest federal role. Jonathan Fisk's thoughtful book examines the diverse range of policy responses across key states and explores the considerable conflict that has emerged between state and local control. This represents an original contribution on fracking governance, with important lessons for the future of American environmental federalism."

—***Barry G. Rabe***, *Gerald R. Ford School of Public Policy, University of Michigan, USA*

The Fracking Debate

The disputes around fracking, and oil and gas policy, follow a long tradition of complicated intergovernmental relationships. Proponents argue that fracking supports new and well-paying jobs, revitalizes state and local economies, and that it can help replace reliance on other fossil fuels. Skeptics and opponents contend that oil and gas production via fracking contaminates air and water resources, causes earthquakes, and can ruin the character of communities. Examining the intergovernmental politics of the first oil and natural gas boom of the 21st century, *The Fracking Debate, Second Edition* offers a holistic understanding of the politics that characterize oil and natural gas operations, including why local governments are challenging their state's preemptive authority, to initiate a larger conversation about improving intergovernmental relationships. Author Jonathan Fisk presents a novel argument about the ways in which local, state, regional, and national approaches to governance of shale gas development can work together to reduce conflict and forward the interests of the communities exposed to development, asking important questions such as:

- What state legal structures govern state-local relations? What state institutions impact and shape oil and gas production? What is the policymaking context in the state?
- What are the costs and benefits of hydraulic fracturing at the national, state, and local levels? How are risks and rewards distributed within states?
- What local policies have challenged the state, and why would local communities challenge the state?

The result is a book that demonstrates that when stakeholders acknowledge their interdependencies and one another's expertise, they create, design, and implement more responsive, strategic, and targeted public policies. *The Fracking Debate, Second Edition* will be required reading for courses on oil and gas policy in the United States, environmental politics, and domestic energy politics, as well as a vital reference for practitioners and policymakers working in these fields.

Jonathan M. Fisk, PhD, is an assistant professor of Political Science at Auburn University, Auburn, Alabama.

American Society for Public Administration
Series in Public Administration & Public Policy
Editor-in-Chief: David H. Rosenbloom, PhD

For a complete list of titles in this series, please visit www.routledge.com

Mission: Throughout its history, ASPA has sought to be true to its founding principles of promoting scholarship and professionalism within the public service. The ASPA Book Series on Public Administration and Public Policy publishes books that increase national and international interest for public administration and which discuss practical or cutting edge topics in engaging ways of interest to practitioners, policy makers, and those concerned with bringing scholarship to the practice of public administration.

Recent Publications

Managing Digital Governance
Issues, Challenges, and Solutions
by Yu-Che Chen

The Future of Disaster Management in the US
Rethinking Legislation, Policy, and Finance
edited by Amy LePore

Case Studies in Disaster Response and Emergency Management, Second Edition
by Nicolas A. Valcik and Paul E. Tracy

Researcher-Policymaker Partnerships
Strategies for Launching and Sustaining Successful Collaborations
by Jenni W. Owen and Anita M. Larson

Advances in E-Governance
Theory and Application of Technological Initiatives
by Anthony Trotta

The Fracking Debate
Intergovernmental Politics of the Oil and Gas Renaissance, Second Edition
by Jonathan M. Fisk

City Planning for the Public Manager
by Nicolas A. Valcik, Teodoro J. Benavides, Todd A. Jordan, and Andrea D. Stigdon

The Fracking Debate

Intergovernmental Politics of the Oil and Gas Renaissance

Second Edition

Jonathan M. Fisk

Routledge
Taylor & Francis Group
NEW YORK AND LONDON

First published 2018
by Routledge
711 Third Avenue, New York, NY 10017

and by Routledge
2 Park Square, Milton Park, Abingdon, Oxon OX14 4RN

Routledge is an imprint of the Taylor & Francis Group, an informa business

Library of Congress Cataloging-in-Publication Data
Names: Fisk, Jonathan M., 1983- author.
Title: The fracking debate : intergovernmental politics of the oil and gas renaissance / by Jonathan M. Fisk.
Description: New York : Routledge, 2017. |
Includes bibliographical references and index.
Identifiers: LCCN 2017009594 | ISBN 9781498742412 (hardback : alk. paper) | ISBN 9781315212715 (ebook)
Subjects: LCSH: Hydraulic fracturing—Environmental aspects—United States. | Hydraulic fracturing—Political aspects—United States. | Energy policy—United States. | Gas industry—United States. | Intergovernmental cooperation—United States.
Classification: LCC TD195.G3 F57 2017 | DDC 622/.3381—dc23
LC record available at https://lccn.loc.gov/2017009594

ISBN: 978-1-4987-4241-2 (hbk)
ISBN: 978-1-315-21271-5 (ebk)

Typeset in Sabon
by Apex CoVantage, LLC

To my grandparents, who inspired my interest in politics; to my parents, my first academic role models; to Melanie, my wife and best friend; and to Noah, my son and inspiration to complete this book

Contents

Detailed Contents

About the Author

Jonathan M. Fisk, PhD, is currently an assistant professor of Political Science at Auburn University, Auburn, Alabama. He holds a Doctorate in Political Science from Colorado State University and a Master's of Public Administration from the University of Kansas. He has served as a Research Associate with the League of Kansas Municipalities. He is also the Faculty Adviser to the Auburn University Student Chapter of the International City and County Management Association, the Secretary of the Alabama Chapter of the American Society for Public Administration, and a board member for the Section of Environment and Natural Resource Administration. His research has appeared in a number of academic outlets, including State and Local Government Review, Society and Natural Resources, Politics and Policy, Review of Policy Research, and Local Environment. At Auburn University, he teaches graduate seminars in Public Personnel Management, Organizational Theory, and Leadership.

Acknowledgments

I would like to acknowledge the substantial investment in energy, time, resources, and support of so many that made this project possible. In particular, I would like to thank the American Society for Public Administration for nurturing and supporting critical practitioner-academic exchanges. I would also like to thank the many people at Taylor and Francis, who assisted me in completing this book. I also owe a special thanks to Laura Stearns and Misha Kydd for their help in completing, promoting, and marketing this book. I would also like to acknowledge those individuals who generously gave me permission to incorporate their work. The project is sufficiently stronger because of your previous work. Finally, I want to acknowledge the two anonymous reviewers, whose suggestions greatly improved this book.

I also owe a debt of gratitude to Charles Davis, Susan Opp, John Straayer, and Tara Shelley, who guided me during my time at Colorado State University—where this project began. I would also like to thank Auburn University, the College of Liberal Arts, and the Department of Political Science, especially the MPA program, for providing me with research support and time needed to complete this project.

Most important, I want to thank my family for all of their support and encouragement. Each of you is owed a tremendous debt of gratitude.

Preface

State-local relationships are complicated. They are not only shaped by formal laws, rules, and court decisions but also by the informal organizational values, norms, and relationships of actors. Subnational relationships are also related to the rhetorical and framing strategies utilized by stakeholders. Existing research has identified these dynamics in a wide variety of policy areas, including American foreign policy, living and minimum wages, transgender rights, and the enforcement of some of the nation's environmental laws (Hobbs 1994; Riverstone-Newell 2012). In this sense, the intergovernmental disputes revolving around fracking and oil and gas policy follow in a long tradition of complicated intergovernmental governing relationships.

Fracking is a core component of the oil and gas industry's toolbox. It, however, is only one piece of the production process, and it is being used as part of the extraction process in thousands of wells across the United States. Proponents argue that oil and gas development via fracking supports new and well-paying jobs, revitalizes state and local economies, and can replace other fossil fuels as the nation's primary baseload electrical power–generating fuel. Skeptics and opponents contend the oil and gas production contaminates air and water resources, causes earthquakes, and can ruin the character of many communities (Davis 2012; Fisk 2016).

This project examines the intergovernmental politics of the first oil and natural gas boom of the 21st century and has two main goals. The first is to offer a more holistic understanding of the intergovernmental politics that characterize oil and natural gas operations, including why local governments are challenging their state's preemptive authority. The second is to begin a larger conversation about improving intergovernmental relationships when it comes to oil and gas production. To accomplish these goals, it is imperative to explore three broad areas as they relate to the intergovernmental management of oil and gas production:

1. What state legal structures govern state-local relations? What state institutions impact and shape oil and gas production? What is the policymaking context in the state?

2. What are the costs and benefits of oil and gas production at the national, state, and local levels? How are risks and rewards distributed within states?
3. What local policies have challenged the state and why would local communities challenge the state?

Chapter 1 addresses the oil and gas debate at large. Throughout the chapter, attention is paid to specific environmental costs and economic benefits in an attempt to provide a more complete understanding of the production process. Chapter 2 pivots to the growing role that federal government plays in onshore oil and gas production, with an explicit emphasis on areas of federal policy that impact state and local governments. Previous work, for example, has noted that the federal government has exempted fracking from a number of federal environmental laws (Davis 2012). Yet, when it comes to regulating fracking on public lands, rail safety, pipelines, and offshore development, federal agencies are active policymaking institutions. Chapter 3 summarizes the legal principles that govern the states and local governments and how each impacts the environment. The chapter also provides a cursory overview of local defiance across the United States and highlights possible reasons why local governments may pushback against state policies that encourage oil and gas extraction.

Chapters 4 through 6 present state level case studies of oil and gas politics in Colorado, Ohio, and Texas. These three chapters dig deeper into intergovernmental relations and attempt to answer the first two questions listed previously. As such, these chapters provide important foundations for understanding local defiance that is presented in Chapters 7 and 8. Each chapter summarizes the legal structures that govern state-local relationships and the institutions that manage oil and gas production. These chapters also address the 'geography' of production and how this distribution shapes the distribution of economic rewards and environmental risks and subsequently the intergovernmental politics of oil and gas.

Chapters 7 and 8 unpack the sources of local defiance in Colorado, Ohio, and Texas. Chapter 7, through a series of short case studies, examines the possible sources of conflicted intergovernmental relationships that have resulted from expanded oil and gas development. This chapter also includes several cities, which have policies that align with their state. Chapter 8 uses statistical models to assess possible factors that may be associated with challenges to the state's preemptive authority. The local governments included in these chapters represent different regions of the country, sets of politics, and experiences with oil and gas development. Even with this variation, they also illustrate a common theme that is present across many states—despite a variety of state regulatory approaches, local defiance is real and because of the complexity related to oil and gas governance, new solutions are needed. Chapter 9 builds on the sources of local concern and identifies best practices ranging from new regulatory/legislative solutions to strictly voluntary programs that may be applied across the country to improve intergovernmental relations.

1 Introducing the Politics of Oil and Gas Development

From those concerned primarily with its forms and expression, to those more concerned with how the governing authority ebbs and flows, scholars have long concerned themselves with the concept, vicissitudes, and displays of political power (Agranoff and McGuire 2001; Bachrach and Baratz 1962). They have also concerned themselves with how power dynamics are at the core of many contemporary intergovernmental challenges. They are evident, for example, when one unit of government compels or restricts another's agenda, during an intergovernmental lawsuit, or when one government lobbies another (Cobb and Elder 1972; Schneider, Jacoby, and Lewis 2011). Questions about power are often evident when a level of government challenges another. Intergovernmental power, as it relates to states and local governments, even when it favors the states, however, is often shared, uncertain, and ephemeral (Long 1949). Anton (1989) adds that because it is split among many stakeholders, tensions and contests are inevitable (Bowman and Kearney 2012, 2011; 2010; Stephens 1974).

Such power dynamism is evident in the actions, goals, resources, processes, symbols, and intergovernmental strategies acted upon or utilized by the actors participating in the fracking-fueled oil and natural gas debate (Betsill and Rabe 2009; Bosso 2005; Davis 2014; Hempel 2009; Lowry 1991; Opp and Saunders 2013; Rabe 2010, 2006, 2004; Stoker 1998; Vig and Kraft 2009; Walker 1995). On one hand, the 'fracking' debate involves questions over whether oil and gas production is harmful or beneficial. On the other hand, it also goes to a fundamental 'place' within the American intergovernmental system and forces stakeholders to consider difficult questions about property rights and local democracy. In other words, the 'boom' has also led to more traditional ones about governance. Who governs? Who should govern? How do they govern? What do they govern? Are there more effective forms of regulation that can assuage the reasonable fears of impacted populations while protecting private property rights, economic growth opportunities, and the jobs created by domestic production? The answers to these questions are critically important, as both states and local governments are needed to meaningfully address contemporary environmental challenges.

The federal government has taken the regulatory lead on issues such as railroads, railcar safety, federal lands management, and interstate oil and gas pipelines. Yet, for many of the core issues of oil and gas, the absence of the federal regulation has enabled a vigorous state-local federalism to emerge. Subnational governments are responsible for managing the placement of wells, infrastructure, the technical aspects of drilling, managing potential impacts to the environment, site security, and emergency management. They are also in charge of addressing many of the quality of life impacts caused by oil and gas production. As a result, such subnational policymakers are the primary recipients of the economic rewards of oil and natural gas and also experience the costs of production firsthand.

With the recent advances in drilling and production practices, the stakes for state and local governments have become even higher as unconventional oil and gas extraction is now taking place in more densely populated suburban and urban communities. In many cases, the economic goals of states and their cities have aligned to favor oil and gas development. This is the case for many cities across the United States. Yet, examples of local defiance are evident from California to New York, although the triggering factors appear to be different. In short, the costs and benefits of the 'boom' have strained the intergovernmental system, lead to acts of local defiance, and call for new and inventive approaches to managing oil and gas production.

This project attempts to understand local defiance and decision-making in the context of the highly fragmented system that governs fracking and oil and gas production (Fisk 2016; McAdam and Boudet 2012). Although local defiance is certainly not the preferred pathway of most local governments, it is and has been an option pursued by a small group of communities. To improve the intergovernmental management of oil and gas production, this project examines local defiance in Colorado, Ohio, and Texas with a three-part process:

1. The applicable state rules, policies, and judicial interpretations of state-local relationships in three different states
2. Specific state factors and implications such as the distribution of costs and the location of reserves/production
3. The local experiences as they relate to oil and gas development and how such experiences have impacted state-local relationships

These states represent three different sets of oil and gas experiences, geographical regions, and political leanings. Additionally, each of these states has examples of intergovernmental tension but has also experimented with inventive policy prescriptions. In short, by incorporating three divergent states, the factors associated with defiance become more generalizable and the lessons learned from them become more transferable to other jurisdictions.

Locating and then explicating the sources of intergovernmental tension are important first steps. Once accomplished, the second and more difficult challenge is to then identify more optimal governing arrangements. Solutions must fit within the fragmented regulatory system that governs oil and gas production. And, they must incorporate the strengths of both state and local levels of government as each brings significant pieces to the regulatory puzzle. Through leveraging each other's strengths, policymakers may design more effective, strategic, and efficient forms of environmental governance.

The Basics of Production and Fracking

Since 1946, operators have vertically drilled and then fracked more than one million oil and natural gas wells, most with little fanfare and few cases of contamination (American Petroleum Institute 2014). Since the mid-2000s, technological advances in the extraction processes, drilling, and the proprietary chemicals used by operators set the stage for the oil and gas boom (Davis 2012). In 2005, this confluence of factors made shale gas production economically viable. Later, similar technologies were applied to tight oil formations. Production from these regions started growing rapidly around 2008. Today, rather than drilling multiple vertical wells, industry (often oilfield service operators) is now able to drill up to 10,000 feet downward and through horizontal drilling follow the mineral deposit (between 4,000 and 10,000 feet) (Dunn 2014). Then, they can frack along the shale or tight oil deposit. Unlike the recent past, contemporary production has aroused the public's attention, elicited its animosity, impacted residents in a multitude of communities, and has complicated state-local relationships.

Twenty-first century 'fracking' is still part of the larger oil and gas production process. This, in turn, can be divided into several stages: pre-drilling, permits/drilling, and post-drilling activities. Although specific requirements depend on the governing jurisdiction, geology, and operator, the 'process' generally fits a rough pattern outlined in Figure 1.1 (Dunn 2014; FracFocus 2015; Pless 2011).

The 'drilling and fracking portions' of the process are displayed in more detail in Figure 1.2.

The various 'steps' provide activists with opportunities to challenge the status quo, to impede development, and to seek out more receptive institutions. It also means that in many instances, no single agency or institution exists that holds a monopoly on regulatory or political power. Equally important, the reality of multiple 'steps' leaves ample room for intergovernmental, inter-agency, and cross-sectoral collaboration so that

Pre-Drilling

- Identifying, mapping, surveying, and inventorying potential sites
- Securing rights of way and leasing
- Mapping of the surface and subsurface formations, impacted fault lines
- Sampling and testing of nearby resources

Permits and Paperwork

- Submit applicable permits to state agencies – construction and drilling are the most common
- Perform any required studies such as wildlife impact assessments, seismic surveys, etc.
- Submit applicable surety and bonding amounts

Drilling

- The well is drilled (includes both vertical and horizontal drilling)
- The well is constructed

Hydraulic Fracturing (Fracking)

- Begins when operators (often an oilfied service company) inject a high-pressure blend of sand (between 1.5 and 6 million pounds), water (may exceed 4 million gallons), and chemicals into the formation (Dunn 2014).
- The unique chemical cocktail is related to the specific company, well type, and geological formation/surrounding ecology (Davis 2012; EPA 2011; Fisk 2013; National Conference of State Legislatures 2010; Pless 2011).
- As pressures inside the well increase, the shale formation eventually 'fractures' and releases the trapped oil or gas. This is the part of oil and gas production that is commonly referred to as 'fracking.'

Post-Production

- Waste Disposal
  - Wastes can be recycled, stored on site, or injected into wastewater disposal wells
- Well capping and plugging

Figure 1.1 The Production Process (in general)

stakeholders may engage in dialogues with one another to prevent and/or solve challenges as they arise. In this way, the intergovernmental system invites conflict but also lays the foundation for new and innovative governing mechanisms that reduce intergovernmental tensions and better respond to citizen demands. See King (2012) for a thorough and comprehensive resource that describes the technical aspects of the fracturing process.

The Oil and Gas Debate

The questions and debates related to hydraulic fracturing (fracking) enable participants to conjure powerful political images, iconic symbols, and language. Opponents claim that fracking poisons the environment

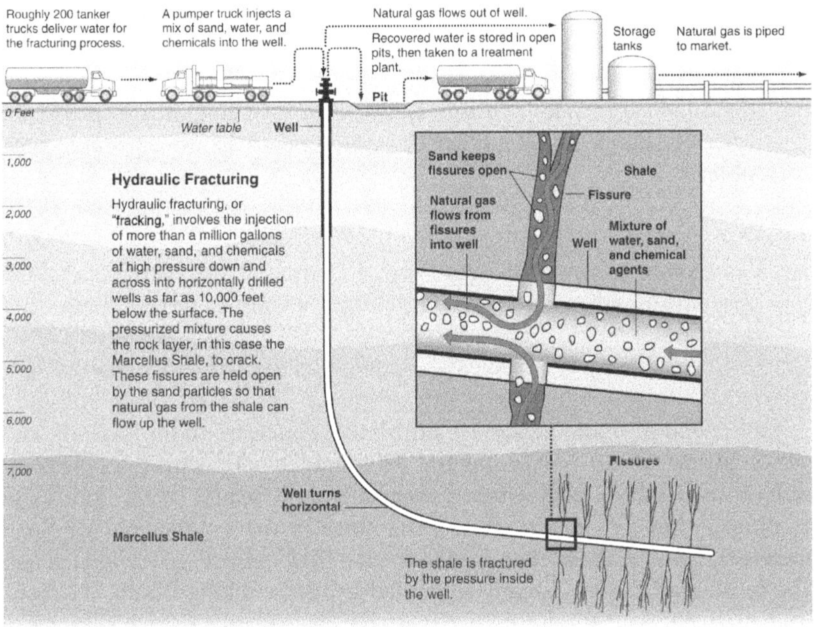

Graphic by Al Granberg

Figure 1.2 The Fracturing Process

Source: Al Granberg/ProPublica (http://wisconsinwatch.org/wpcontent/uploads/2011/07/ProPublica_hydrofracking_graphic.pdf)

and that unfair collusion between state officials and industry squashes dissent and dismisses grave public health concerns. Proponents argue that oil and gas production via fracking is an economic lifeline for many financially distressed communities and states, and that it moves the United States closer towards achieving energy security and independence. Best management practices often focused on technical aspects of drilling and fracking, they add, are cost effective and may be used to mitigate many environmental risks. These contrasting viewpoints are exemplified by the recent debate in several northeastern states: Pennsylvania, where oil and gas operations are ongoing; and in New York State, where production via fracking is banned; and in Maryland, where the practice is narrowly permitted (Phillips 2014; Wagner 2014).

- Howard Zucker, the acting commissioner of the New York State Department of Health, described the environmental and health impacts of fracking as "Would I let my family live in a community with fracking?' The answer is no."

  (Zucker as quoted by Rochester Business Journal 2014)

- Former Democratic Governor Martin O'Malley described fracking's risks as manageable, "Provided all the recommended best practices are followed and the State (Maryland) is able to rigorously monitor and enforce compliance, the risks of Marcellus Shale development can be managed to an acceptable level" (O'Malley as quoted by Phillips 2014). The state's current governor, Larry Hogan, a Republican, described fracking as having the potential to be an economic gold mine.

  (Wagner and Clement 2015)

- Pennsylvania Governor Tom Wolf, a Democrat, also offered a telling observation when it comes to fracking: "I want to have my cake and eat it too."

  (Wolf as quoted by Colaneri 2014)

Embedded in these quotes are normative concerns about fracking and observations about environmental governance. For some, the lack of federal oversight is problematic, especially as it relates to air and water pollution that can easily cross state lines. Others advocate for local control. They suggest that, because city and county governments are the most likely to be directly impacted by the fracking boom, its costs, and its benefits, they should assume a more assertive role in mitigating its impacts, typically through their zoning, land use, and public safety authority. Finally for some, state governments are the optimal loci of authority. States, they argue, have traditionally regulated the industry, and as such, have the requisite expertise to balance the costs of development with its benefits.

States and local governments have confronted the tangible and intangible risks and rewards of unconventional oil and gas production. Whether real or perceived, how state and local stakeholders interpret and understand the costs and benefits of oil and natural gas development matters greatly and is summarized in the remainder of Chapter 1. If state and local governments believe that benefits are high with manageable costs, local communities may be more willing to cooperate with state regulations and absorb, plan for, and mitigate possible risks. Conversely, if the costs are high, local governments may view the risks as unacceptable and engage in acts of local defiance. Thus, local resistance does not exist in a political vacuum and functions as a response to several interrelated pieces:

1. State regulations or policies (i.e., what are local governments defying or challenging)
2. Distribution of costs and benefits (i.e., who is absorbing the risks and rewards associated with oil and natural gas production)

3. Local preferences and characteristics (i.e., what about the local government leads to a misalignment between community's values and that of the state)

Most local governments work within their state's established regulatory framework. However, a small number of communities have become a receptive venue for actors opposed to oil and gas production. As a result, city council chambers and the city ballot initiatives have become focal points for ideological forces engaged in defining who benefits/wins from oil and gas development and who loses/absorbs the costs of production.

The Benefits of Oil and Gas Development

As an energy source, unconventional oil and natural gas development offers a number of advantages. First, it supports thousands of direct and indirect jobs, many of which pay above regional or state averages. Second, through severance taxes, impact fees, and other revenue streams, hydrocarbon production sends millions of dollars in revenues to state and local budgets and to private citizens (Davis 2012). Third, natural gas (to a lesser degree oil) is considered a baseload-generating fuel, meaning that it can produce reliable and consistent electrical power. Fourth, oil and natural gas are used in a variety of applications and sectors. They are found in fertilizers, pharmaceuticals, paints, cosmetics, toys, sports and recreational equipment, medical devices, and electronics (American Natural Gas Alliance 2015). Additionally, oil (and now natural gas) can be liquefied and used to power vehicles or exported. A fifth set of benefits, according to supporters, relates to climate change. Relative to other fossil fuels, natural gas, when combusted, releases less carbon dioxide compared with coal. Its use, especially as a replacement for coal, may be able to help slow down the effects of anthropogenic climate change (EPA 2011; Union of Concerned Scientists N.d.b). Finally, fracking has reduced energy and transportation costs for consumers.

Revenue Benefits

Production generates billions of dollars in revenues for state and local governments. In fiscal year (FY) 2010, for example, oil and gas economics contributed more than $18 billion to state and local economies (Pless 2012). Many of these positive economic indicators have continued through 2015, including more than $255 billion in economic output (from extraction) alone in 2015, as shown in Tables 1.1 and 1.2.

Table 1.1 Industry's Economic Output (in billions)

	2008	*2009*	*2010*	*2011*	*2012*	*2013*	*2014*	*2015*
Mining	592.4	380.4	463.3	556.1	570.2	619.5	681.6	449.8
Oil and gas extraction	395.4	223.6	289.7	343.7	341.3	396.7	436.8	255.4
Mining, except oil and gas	116.5	104.4	114.7	135.9	133.5	125.8	133.9	117.4
Support activities (for mining)	80.5	52.4	58.9	76.5	95.4	97	111	77

Source: Bureau of Economic Analysis (2016)

The taxes and fees collected during oil and gas production also help fund numerous state and local governments. In fact, some governments depend on industry-related revenues for large portions of their general fund budget and to support their rainy day accounts. Other jurisdictions are less reliant on oil and gas dollars. Typically, these governments favor using industry dollars to offset revenues lost from other income and property tax cuts/declines or to cover the costs of development (Davis 2012; Warner and Shapiro 2013). Oil and gas finances also support other one-time budgetary expenditures. Table 1.3 includes a variety of potential revenue tools/instruments available to state and local governments. The ability to levy and collect each is defined in state regulations and local ordinances.

Currently, more than 35 states collect some sort of fracking-related severance tax, sales/property tax, or impact fee (National Association of State Budget Officers 2015). States and localities, however, vary in the rates they charge, the generosity of available tax exemptions and incentives, the factors used to calculate industry's tax liability, and the degree to which states share revenues with city or county governments. Determining the price at which the taxes are levied adds further complexity for regulators. Boersma and Johnson (2012) explain that changes in demand, the stage of production, pipeline costs, environmental regulations, fees, and even export markets (assuming no requisite increase in supply) further impact the price of oil or gas. A brief snapshot of how state oil and gas taxes vary is offered here:

- In Alaska, the state's Petroleum Profits Tax funds a large portion of the state's overall budget and is a significant part of the state's economy (Brown and Yücel 2013). The severance tax rate is based on the net value of oil and gas and can range from 25 percent to 50 percent. (Pless 2012a)

Table 1.2 Employment Data

Industry	*Employment*							*Output*				
	Thousands of Jobs			*Change*		*Compound Annual Rate of Change*		*Billions of Chained 2009 Dollars*			*Compound Annual Rate of Change*	
	2004	*2014*	*2024*	*2004–2014*	*2014–2024*	*2004–2014*	*2014–2024*	*2004*	*2014*	*2024*	*2004–2014*	*2014–2024*
Mining	523.2	843.8	924.0	320.6	80.2	4.9	0.9	372.2	459.1	556.1	2.1	1.9
Oil and gas extraction	123.4	197.9	219.4	74.5	21.5	4.8	1.0	207.0	266.5	316.6	2.6	1.7

Source: BLS (2016)

Table 1.3 Revenue Tools and the Shale Boom

Tool	*Explanation*
Severance taxes	Taxes that are assessed once the gas or oil is 'severed' from the ground—can be based on the volume of gas/oil produced, number of wells, value of the oil/gas, etc.
Impact fees	Fees are revenues typically collected to mitigate the costs of development such as road deterioration, park degradation, etc.
Sales taxes	Sales taxes are levied on the value of sales for oil and gas equipment and other indirect sales (varies by state). Direct and indirect collections can be subject to exemptions.
Property taxes	Property taxes are taxes levied against the value of property owned by the oil and gas industry and other indirect revenues received (varies by state along with potential exemptions).
Administrative fees	State and local governments often collect permit fees, etc., that cover the costs of development or help fund the applicable regulatory agency.
Lease payments (gas or water)	Lease payments are rates, fees, etc., that are negotiated in the lease terms and accrue to the owner, which can include states/ local governments and private citizens.
In-kind donations	Donations made to applicable state/local governments or non-profit/foundations.
Fees/fee for service	Payments for services rendered are considered fees for service, such as filing paperwork.
Royalties	Monies paid to landowners, as agreed in the lease agreement.

Source: Davis (2012), Minor (2014), Raimi and Newell (2014)

- Utah charges developers a severance tax, with the rate dependent on the value of natural gas. The state collects "3 percent of the value up to and including the first $1.50 per million cubic feet for gas" (Pless 2012a). This rate increases to 5 percent of the value, from $1.51 and above per MMcf. The state also charges 4 percent severance tax on the taxable value of natural gas liquids and a .002 percent conservation fee (Pless 2012a). The cumulative financial impact to the state's general fund was approximately $65 million in FY 2012.

  (Salt Lake 2012)

- Through its 11.5 percent severance tax, North Dakota collected nearly $2 billion in FY 2011. In FY 2012, severance tax collections surpassed $3 billion. Before the discovery of the Bakken Formation, the state collected less than $100,000 in severance taxes.

  (Phillips et al. 2013; Pless 2012a)

- The Texas budget benefits from a 7.5 percent severance tax placed on the market value of gas (Pless 2012a). In 2011, gas-related revenues totaled $2.7 billion (Prah 2013). For FY 2014–2015, gas production taxes were valued at approximately $3.5 billion and even with a recent dip in prices, FY 2016–2017 collections were estimated at $3.23 billion.

(Nichols 2015)

- Colorado's severance tax rate ranges from 2 to 5 percent based on the gross income generated from oil, gas, carbon dioxide, and coal-bed methane extraction. Despite having one of the nation's lower effective severance tax rates (more in Chapter 4), the tax generates millions for state and local lawmakers. In FY 2014–2015, severance tax revenue surpassed $369 million, a $100 million increase from FY 2013–2014. State officials expect that the dive in prices will contribute to a decline in severance tax revenues in FY 2016–2017.

  (Denver Post Editorial Board 2015)

- Illinois also exemplifies the economic allure of domestic oil and gas production. Before the downturn in oil prices, the state was primed to see unconventional oil and gas production. An analysis conducted by Headwater Economics for the Chicago Tribune, which assumed a price of $105 a barrel of oil over a 10-year period and that the state's tax rates/structure would remain stable, concluded that the state would have collected around $904,046 in severance taxes per well. At that price, Headwater economists anticipated that drilling would be profitable (Wernau 2015).

Other states are listed in Table 1.4.

Table 1.4 Selected States and Severance Tax—2013

	Total Taxes (in thousands)	*Severance Taxes (in thousands, including coal and other minerals)*	*Severance Tax Proportion of Total Taxes*
Alaska	5,132,811	4,016,966	78.26%
Louisiana	9,223,829	834,116	9.04%
Montana	2,644,610	282,356	10.68%
Nevada	7,026,626	290,448	4.13%
New Mexico	5,201,576	713,998	13.73%
North Dakota	5,298,770	2,457,530	46.38%
Oklahoma	8,892,503	515,981	5.80%
Texas	51,714,295	4,647,848	8.99%
West Virginia	5,378,122	608,371	11.31%
Wyoming	2,186,054	867,933	39.70%
Colorado	11,245,662	147,732	1.31%

Source: U.S. Census (2014, 2013)

Given the economic potential of oil and gas development, state lawmakers from across the United States have actively debated their severance tax in recent years. Table 1.5 offers a snapshot and a brief summary of changes.

Table 1.5 Significant Severance Tax Changes

State	*Year*	*Legislation*	*Description*
Alaska	2013	SB 21	Changes tax rate to 35% of the production value of oil and gas.
Alaska	2007	HB 2001	Increases tax rate to 25%. Known as Alaska's Clear and Equitable Share (ACES).
Alaska	2006	HB 3001	Establishes tax rate at 20% of net profits. Known as Petroleum Profits Tax (PPT).
Colorado	2014	SB 14–154	Allocates severance tax operational fund revenue to the wildfire preparedness fund.
Colorado	2014	HB 14–1333	Transfers severance tax perpetual base fund to water conservation board construction fund.
Colorado	2013	HB 13–1057	Allocates severance tax trust fund revenue to geological survey and avalanche information center.
Colorado	2012	HB 12–1315	Allocates revenues to innovative energy fund and local government severance tax fund.
Colorado	2012	SB 12S-002	Transfers revenue to the Colorado water conservation board for reservoir projects.
Colorado	2010	HB 10–1250	Allocates perpetual base account revenue to the Colorado water conservation board construction fund for Animas–La Plata Project Water.
Colorado	2009	SB 09–165	Allocates perpetual base account to small communities' water and wastewater grant funds.
Colorado	2009	HB 09–1199	Transfers revenue to healthy forests and vibrant communities fund and wildland–urban interface training fund.
Idaho	2012	HB 379	Amends two severance taxes, changing tax rate to 2.5% of market value of oil and gas.
Illinois	2013	SB 1715	Known as the Illinois Hydraulic Fracturing Tax Act: establishes Illinois severance tax.
Louisiana	2009	SB 765	Allocates portion of revenues to the Atchafalaya Basin conservation fund.

State	*Year*	*Legislation*	*Description*
Mississippi	2013	HB 1698	Adds tax rate of 1.3% of value of oil and gas produced from horizontally drilled wells.
North Dakota	2013	HB 1278	Allocates portion of production tax revenues to newly created outdoor heritage fund.
North Dakota	2013	HB 1358	Alters revenue allocation to counties and hub cities.
North Dakota	2013	SB 2014	Allocates oil extraction development funds to newly created energy conservation grant fund.
North Dakota	2010	HCR 3054 and Ballot measure 1	Creates the Legacy Fund and allocates 30% of oil and gas revenues to it.
Oklahoma	2013	HB 2562	Changes rate of Oklahoma's gross production tax on oil and natural gas to 2 percent for the first 36 months of production.
Pennsylvania	2012	Act 13	Creates Pennsylvania impact fee on oil or gas wells produced within the state.
Utah	2012	Ballot measure, Amendment A	Requires specific allocation formula of severance tax revenues to the permanent state trust fund.
Utah	2008	Ballot measure, Amendment B	Permits legislature to direct money or other assets given to the permanent state trust fund under any provision under law.
West Virginia	2014	SB 461	Creates West Virginia Future Fund and allocates 3% of the severance tax revenue to fund.

Source: Courtesy of Rabe and Hampton (2015)

Local governments have also benefited from the shale oil and gas boom through new tax collections, intergovernmental transfers, and other financial instruments outlined in Table 1.3. County impact fees for several Pennsylvania counties located atop the Marcellus Shale exceeded $4 million in recent years. In FY 2012, Colorado's local governments and other local programs (through the Colorado Department of Local Affairs) received slightly more than $60 million from production. Barnett Shale counties, such as Johnson County, Texas, reported that the increases in property values, additional tax revenues, and lease payments were enough to cover the additional expenses associated with wear and tear on the county's infrastructure. In Fort Worth, Texas, economic numbers were even more positive:

- Tax collections from natural gas developers have grown from $1.5 million to $8.5 million since the development of the Barnett Shale began around 2008.

  (Raimi and Newell 2014)

- Seven percent of total retail sales in 2012 were related to mining, quarrying, or to oil and gas extraction.

  (Raimi and Newell 2014)

- Costs (nearly $1 million) associated with the creation of a new administrative division to monitor and enforce the city's gas policies were covered by a permit fee charged on wells located within the city.

  (Raimi and Newell 2014)

Weber, Burnett, and Xiarchos (2016) also found that oil and gas development has been a net economic positive for local governments. Focusing on the Barnett Shale economics in 2013, they compared shale ZIP codes with non-shale ZIP codes and noted the following:

- Shale ZIP codes reported 9 percent advantage in housing values compared with non-shale ZIP codes.
- Shale ZIP codes had higher revenues for schools compared with non-shale ZIP codes.
- Shale ZIP codes reported a 9 percent drop in property tax millage rates imposed by local school districts compared with non-shale ZIP codes.

A 2016 cost-benefit analysis also determined that oil and gas development has been a net positive for communities and households. Bartik et al.'s (2016) analysis concluded that 'fracking' has a number of favorable economic outputs that outweigh its associated costs. Favorable economic indicators include: a 6 percent rise in household income, employment growth of 10 percent, and increased home values by approximately 6 percent, while observing that fracking 'costs' the average household around $1,000 to $1,600 per year in terms of decreasing quality of life (while excluding household income increases). They did note that their results were subject to significant variability across producing regions.

Employment Benefits

The ability to create and/or support high-paying jobs is another economic benefit of unconventional production. Operators in the Pennsylvanian and West Virginian portions of the Marcellus Shale, for example, created nearly 57,000 new jobs in 2011 (The Post-Standard 2011). In fact, according to some estimates, 90% of Pennsylvania's job gains reported between 2005 and 2015 were due to growth in the state's oil and gas sectors (DePillis 2015). Supporters of Ohio's natural gas industry are also citing that the industry generates thousands of jobs (Ohio Department of Job and Family Services 2016). In Colorado and Texas, the oil and gas industry employs hundreds of thousands of workers (Haythorn 2013; Noon 2013). Industry jobs also pay well. In New Mexico, for example, the average employee earns slightly less than $40,000. By comparison, the average oil and gas job in New Mexico pays near $80,000 (Noon 2013; Warner and Shapiro 2013).

The national job picture reflects the optimism found in many statehouses and the industry's inherent volatility (Figure 1.3 and Table 1.6). However, industry's ability to support employment is related to a number of factors such as prevailing commodity prices (high or low), geological characteristics, available infrastructure to transport hydrocarbons (pipelines), the 'local' talent pool, previous costs incurred during well construction, and whether wells can produce more valuable ultra-light crude oil or natural-gas liquids.

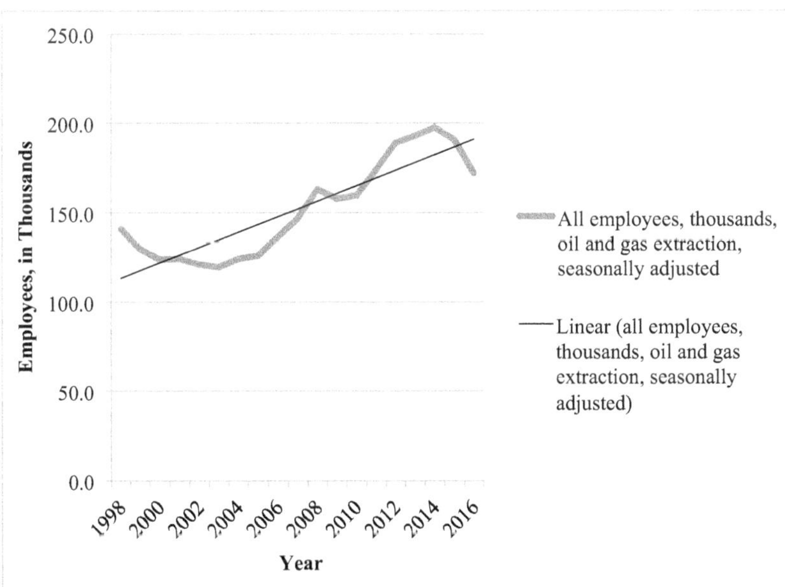

Figure 1.3 Oil and Gas Extraction Employment (seasonally adjusted)

Source: BLS (2016)

Table 1.6 All Employees in Support Activities for Oil and Gas Operations

Year/Month	*Jobs (in thousands)*
1998 Dec	104.5
1999 Dec	99.1
2000 Dec	116.0
2001 Dec	124.1
2002 Dec	122.4
2003 Dec	126.2
2004 Dec	135.9
2005 Dec	155.1
2006 Dec	183.9
2007 Dec	202.3
2008 Dec	221.4
2009 Dec	184.8
2010 Dec	219.4
2011 Dec	267.8
2012 Dec	286.8
2013 Dec	307.6
2014 Dec	334.4
2015 Dec	260.4
2016 Sep (preliminary as of November 2016)	209.9

Source: BLS (2016a)

Oil- and gas-related employment gains extend to non-producing states as well (Sider 2014). A well drilled in 2015, for example, likely required more than 2,000 tons of sand. As such, states with deposits of silica have also benefited from the 'fracking'-fueled renaissance. Wisconsin, before the gas boom, was home to only a few industrial sand producers. By 2015, the state had approximately 100 frack sand operations, which in turn, supported about 20,000 jobs (Ivey 2013; Redden 2013). Opponents, however, warn that industrial frack sand operations are also subject to the boom and bust cycles endemic to any other commodity (Brasier et al. 2013; Gazette Editorial Board 2014).

Energy and National Security Benefits

According to fracking's supporters, unconventional oil and gas is a game changer for America's energy portfolio (Davis 2012). American dependency on foreign oil, they argue, peaked in 2005 with an average of 12.5 million barrels of oil being imported per day (Mb/d). By 2011, this number, largely because of production in areas made possible by horizontal drilling, had dropped to 8.4 million barrels of oil, its lowest level since 1995. Domestic energy production improves national and energy security because it takes place within the United States (i.e., in regions without the geopolitical instability that plague many of the world's oil-producing regions). Stable production creates fewer energy price fluctuations, which enables utilities and other stakeholders to develop longer-term energy forecasts, enter into long-term energy trading contracts, and engage in long-term investment planning (Garfield 2015; Garfield 2015a; Nerurkar 2012; Tomain and Cudahy 2004).

Environmental Benefits

Oil and gas supporters contend that production yields environmental benefits, especially when best management practices are in place (EIA 2016, EIA 2013; Union of Concerned Scientists N.d.). Newly built gas plants, for example, emit between 50 to 60 percent less carbon dioxide compared with newly constructed coal-fired plants (King 2015a). Other near-term environmental benefits include reductions in hazardous air pollutants from coal plants, especially as gas plants replace coal-fired power plants. Industry is also identifying ways to reduce its methane (a potent greenhouse gas) emissions. In 2014, for example, the oil and gas industry emitted approximately 73 million metric tons CO2e (Carbon Dioxide Equivalent) compared with 77 million metric tons CO2e in 2013 (King 2015a). It should be noted that industry's methane 'footprint' is still uncertain and is often specific to individual wells.

Other environmental benefits relate to operators' ability to convert hydrocarbons into fuels used by public transportation. Today, municipal vehicles (cars and buses) and even some taxi operators have turned to liquefied natural gas and have reduced their tailpipe emissions by 15 to 20 percent compared with traditional gasoline (Union of Concerned Scientists N.d.a). In the future, industry expects to be able to process natural gas into pure hydrogen for fuel cell vehicles (Roberts 2004).

Consumer Benefits

The production and consumption of American energy translates into benefits for consumers (EIA 2015; Kuckro 2013). Because much of the infrastructure needed to facilitate expanded gas production is already in place, there are fewer capital costs associated with expanded gas production compared with other renewable sources. A quarter of homes, for example, already rely on natural gas for electricity. As such, natural gas produced largely from shale could account for nearly half of the needed increases in baseload generating power (Union of Concerned Scientists N.d.b).

By increasing the supply of oil and gas, fracking has placed downward price pressures on residential gas prices, as shown in Figure 1.4 and Table 1.7.

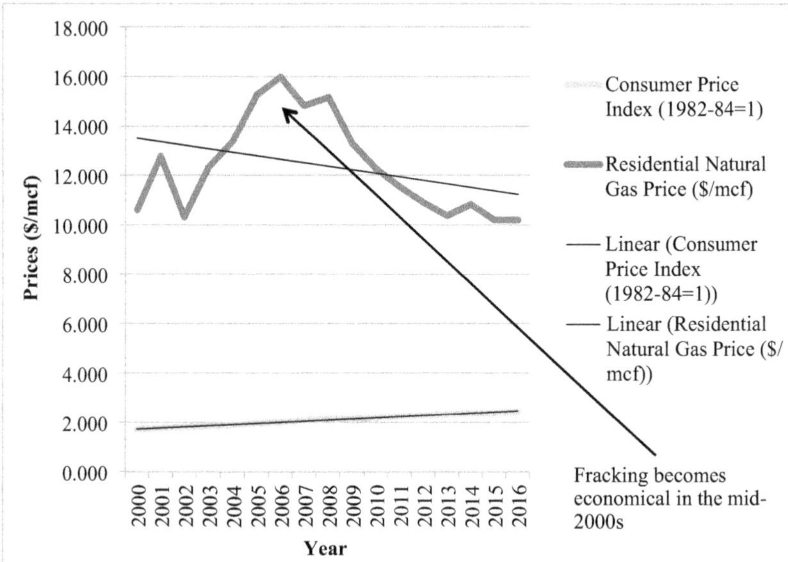

Figure 1.4 Natural Gas Prices
Source: EIA (2016e)

Table 1.7 Energy Prices

Average Cost of Electrical Generation for Natural Gas (indexed to Jan 2008 as value)	
Month	*Cost*
Jan-08	0
Jul-08	3.43
Jan-09	–1.84
Jul-09	–3.96
Jan-10	–1.53
Jul-10	–3.09
Jan-11	–2.89
Jul-11	–3.31

Average Cost of Electrical Generation for Natural Gas (indexed to Jan 2008 as value)

Month	*Cost*
Jan-12	–4.62
Jul-12	–4.88
Jan-13	–3.9
Jul-13	–4.08
Jan-14	–1.19
Jul-14	–3.82
Jan-15	–4.16
Jul-15	–5.17
Jan-16	–5.28
Jul-16	–5.32

Source: www.eia.gov/electricity/data/browser/#/topic/14?agg=1,0,2&fuel=1&geo=vvvvvvvvvvvvo&sec=o3o&linechart=ELEC.COST.NG-US-99.M&columnchart=ELEC.COST.NG-US-99.M&map=ELEC.COST.NG-US-99.M&freq=M&start=200801&end=201608&chartindexed=2&ctype=linechart&ltype=pin&rtype=s&pin=&rse=0&maptype=0

Domestic oil production has also contributed to lower consumer prices. During the 2016 summer driving season, for example, the average gallon of fuel was around $2.04/gallon, a 60 cent decrease compared with 2015 (EIA 2016d). A study, supported by the American Petroleum Institute, concluded that the increase in domestic oil via fracking has brought the price per barrel oil prices down between $12 and $40 in 2013 (ICF International Associates 2014).

Benefits Summary

Oil and gas production creates a number of advantages for state and local governments. Its proponents cite optimistic job creation numbers and rosy revenue forecasts as reasons for subnational governments to establish (or maintain) industry-friendly and centralized regulatory environments. Additionally, they explain that the energy produced via fracking reduces the need for America to import costly oil from unfriendly countries and that it saves consumers billions of dollars at the pump, which can then be spent elsewhere (Goad 2014). Finally, they note that natural gas is a cleaner fuel, especially when compared with coal, and is a key piece in bridging America's growing energy appetite with a clean energy future (Boersma and Johnson 2012; Davis 2012).

Production

The combination of benefits along with the technological advancements associated with fracturing and drilling has led to a resurgence of domestic oil and gas production since the mid-2000s, as shown in Figures 1.5 and 1.6:

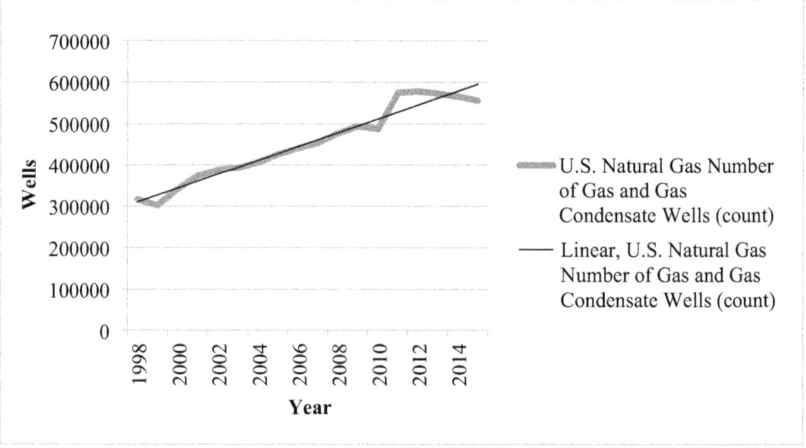

Figure 1.5 Natural Gas Wells
Source: EIA (2014, 2014a, 2014b)

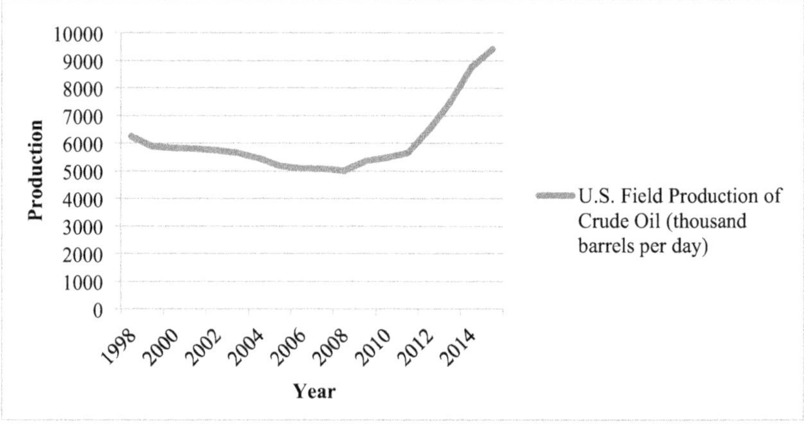

Figure 1.6 Oil Production
Source: EIA (2014c)

Table 1.8 Gas Consumption, by end user (in million cubic feet)

	2010	*2011*	*2012*	*2013*	*2014*	*2015*
Volumes delivered to consumers	22,127,046	22,467,053	23,411,423	23,838,925	24,381,082	25,053,530
Residential	4,782,412	4,713,777	4,149,519	4,897,372	5,087,471	4,609,670
Commercial	3,102,593	3,155,319	2,894,926	3,295,301	3,466,308	3,198,797
Industrial	6,826,192	6,994,120	7,226,215	7,425,452	7,646,039	7,534,589
Vehicle fuel	28,664	29,974	29,970	30,045	35,282	39,379
Electric power	7,387,184	7,573,863	9,110,793	8,190,756	8,145,982	9,671,095

Source: EIA (2016a)

Disadvantages

As an energy source, opponents point to a number of pernicious threats to the environment and to a number of serious disruptions to neighborhood aesthetics and quality of life. Operators, they charge, contaminate air and water resources with a variety of radioactive and dangerous chemicals that are tied to cancer and other poor health indicators. Second, it consumes large amounts of resources, such as water, which can exacerbate management challenges. Third, like other natural resources and commodities, fracking can lead to rapid boom and bust cycles (Detrow 2013; EPA 2015; Goldenberg 2014). During boom times, they also argue, and cite anecdotal evidence, that impacted areas report (residents also perceive) increases in sex/property crimes and drug and alcohol abuse (Jacquet 2014). Growth during the boom is also more likely to be uncoordinated and unplanned, which when development ends, can become blighted and abandoned (Jacquet 2014). Finally, opponents also contend that fracking lowers the quality of life of impacted communities in several ways. In some jurisdictions, it has contributed to deteriorating street and highway infrastructure, ruined a community's visual aesthetics, and disrupted a neighborhood's way of life and peace. It can also increase dust and noise contamination, heighten the risk of man-made earthquakes (tied to wastewater disposal wells), and lead to traffic and railway congestion. Nearby fracking operations may also harm residential property values (Jacquet 2014).

Food and Fracking

Oil and natural gas production requires millions of gallons of water (the industry has explored recycling water), which can complicate food production and has limited water availability in a small number of towns. Colorado's farmers and environmentalists, for example, have expressed concern that energy companies are able to outbid farmers for the state's limited water resources (Jaffe 2014d). In a 2012 New York Times article, Peter V. Anderson, a farmer located on Colorado's Eastern Plains, described how farmers and frackers occupy two different playing fields, "I don't think in reality that the farmer can compete with the oil and gas companies for that water . . . their return is a hell of a lot better than ours" (Healy 2012, NP). Before the boom, Colorado's farmers and ranchers leased excess water from sellers (often cities) for approximately $30 for an acre-foot of water, the equivalent of about 326,000 gallons. Oil and gas companies, however, have offered cities and other sellers between $1,000 and $2,000 for an equivalent amount of water. For many local governments, additional leasing monies (for wells and water) is a new and important revenue stream, but farmers warn that they cannot outbid the industry (Healy 2012).

Like the food produced by farmers, oil can also be shipped via the nation's railways. This increase in traffic has congested the nation's railways as rail supply has failed to keep pace with demand. John Miller, of BNSF Railway, attributed the congestion to "the burgeoning oil and coal sectors . . . even though they use different types of railcars" (Miller as quoted by Husted 2014). He added that BNSR is disappointed about the slowdown, but that delays are impacting all sectors. In 2007, the railroads accommodated 6,000 railcars full of oil; by 2013 the industry was relying on more than 400,000 railcars to ship oil to processing centers. BSNF has announced plans to invest more than a billion dollars in railway expansion, purchasing new railcars and hiring several thousand new people to accommodate the growth (Husted 2014).

Environmental Costs

Opponents argue that oil and gas production is responsible for a number of threats to public health and environmental quality (Davis 2012; Jacquet 2012). Critics have linked the practice to air pollution such as elevated levels of particulate emissions, volatile organic compounds (VOCs), and greenhouse gases. Methane emissions are particularly problematic, because methane is a potent greenhouse gas (King 2015a). The process of production may also threaten drinking water (surface and groundwater) quality through leaks, stray methane, spills, and well blowouts (Jacquet 2014). Table 1.9, based on data compiled by the anti-fracking advocacy

Table 1.9 Select Environmental Impacts of Fracking

Impact	Number
Wells drilled	82,000
Wastewater produced in 2012 (billions of gallons)	280
Water used (billions of gallons)	250
Chemicals used since 2005 (billions of gallons)	2
Air pollution (tons)	450,000
Greenhouse gas emissions (million metric tons or equivalent)	100
Acres damaged	360,000
Induced seismic events in 2014 (USGS 2016, 2014)	659 M3 or larger
Sand used in 2014 (billions of pounds) (Cockerham 2014)	95
Number of spills, blowouts, leaks in 2013 (in 15 top states for onshore oil and gas activity) (Soraghan 2014)	7,662
Annual number of birds lost in oil field production and in other wastewater disposal areas (Federal Wildlife Service N.d.)	500,000 to 1 million

Source: Ridlington and Rumpler (2013)

group, Environment America, quantifies many of fracking's environmental impacts:

Public Health and Air Quality

Environmentalists contend that operators release dangerous air pollutants that can damage nervous, cell, immune, gastrointestinal, respiratory, eyes, skin, and cardiovascular systems (Colburn et al. 2011; Kelly 2012). There are risks of local air pollutants near production facilities, but it should be noted that these are typically much smaller compared with the same types of pollutants emitted by coal-fired power plants:

- A Colorado study showed a relationship between natural gas development and increases in short-term exposure to compounds that have neurological and respiratory implications for residents within a half-mile of a well while it is being completed. The study also found an elevated risk for benzene exposure.

  (McKenzie et al. 2012)

- A Fort Worth, Texas, study found no serious public health risks for individuals living more than 600 feet from gas operations.
- In 2011, sites located near Pinedale, Wyoming, reported ozone pollution levels at 124 parts per billion (ppb). This level is approximately 67 percent higher than the maximum daily limit established by the EPA.

  (Kuster 2012)

In general, emissions related to oil and gas production account for a small majority of industry's environmental footprint, although, when wells are aggregated across an entire city or region, air quality can become more significant. It should be noted that a small percentage of wells have significant leaks (Jackson et al. 2014).

Water Pollution

Like other extractive methodologies, oil and natural gas production can impact surface and groundwater quality. Contamination may occur at multiple points during construction and operation, and post-drilling even though contamination via fracking fluids tends to be a rare event:

- When wells are constructed, stormwater runoff may enter into municipal stormwater systems and other nearby bodies of water. Under the Energy Policy Act of 2005, National Pollutant Discharge permits are not required for oil and gas facilities.

- During operational phases, spills, poor well integrity, and casing failures may lead to fluids contaminating areas near the well, including groundwater.
- The improper storage or disposal of wastes can also lead to contamination (Rabe and Borick 2013; Wiseman 2009). Research, for example, that examined water quality in waters located downstream from a wastewater facility in Pennsylvania found unusually high levels of radioactive elements (bromide and radium), salts, and metallic compounds.

(Warner et al. 2013; EPA 2016)

The question of the relationship between oil and gas drilling and water contamination remains hotly contested (Vengosh et al. 2014). The answer often depends on whether researchers are looking for methane migration or fluid cross-contamination. In general, methane migration has been documented in many wells, whereas fluid cross-contamination (i.e., the iconic images of tap water igniting) is a considerably more infrequent event. Nonetheless, several of the studies that follow find evidence that both proponents and opponents may cite as they engage in the debates swirling around contemporary oil and gas production and operations (Jackson et al. 2014):

- Jackson et al. (2014) examined the possible contamination between oil and gas development and drinking-water wells for 141 homes located above the Marcellus Shale (Pennsylvania). Data showed evidence of elevated natural gas concentrations in the wells, which they attributed to casing and cementing failures.
- Sloto (2013) measured possible cross-contamination for 20 homes in Sullivan County (Pennsylvania). Of the 20 homes, the researcher noted evidence of cross-contamination in the wells that supplied drinking water to two homes.
- A study of Pavillion, Wyoming, detected high levels of benzene and other chemicals, which are frequently used during fracking operations, near underground sources of drinking water (DiGiulio and Jackson 2016). The validity of these findings was later challenged by Connor et al. (2016).
- Fontenot et al. (2013) sampled 100 drinking-water wells located within the Barnett Shale play. The study found "higher levels of arsenic, selenium, strontium, and total dissolved solids (TDS) in water wells <3 km from shale-gas wells." The researchers cautioned, however, that follow-up studies were needed before any final determination could be made as to the causal relationship between fracking and contamination.

(See also Jackson et al. 2014)

In 2016, the EPA also released a report documenting the relationship between contamination and fracking. The agency found that possible drinking-water impacts occur at multiple points during production (including but not limited to fracking). More severe environmental impacts may occur at the following times:

- During or in regions with low water availability; risks are acute during or in regions that are reporting that groundwater resources are declining
- Large-volume spills or spills with a high concentration of fracking fluid
- Wells with poor or inadequate construction
- The injection of fluids into groundwater resources (limited in many states)
- Poor disposal practices or the storage of wastewater in unlined pits (limited in some states)

Despite identifying the points of greatest risk, the EPA report noted that "because of these data gaps and uncertainties, as well as others described in the assessment, it was not possible to fully characterize the severity of impacts, nor was it possible to calculate or estimate the national frequency of impacts on drinking water resources from activities in the hydraulic fracturing water cycle" (EPA 2016a). The report also dropped language that described fracking's impact on drinking water as "widespread" and "systemic."

State records also suggest a mixed bag for stakeholders engaged in conversations about managing oil and gas production. Pennsylvania data indicate that between 3 percent and 6 percent of Pennsylvanian wells failed during their first three years (Jackson et al. 2014). Yet, in Arkansas, regulators have not recorded any examples of drinking water contamination related to drilling activities (Jackson et al. 2014). Other examples are given as follows (Begos 2014):

- Pennsylvania officials confirmed approximately 106 cases of water-well contamination since 2005. In that same time frame, gas firms have drilled more than 5,000 new wells across Pennsylvania.

  (Begos 2014)

- Between 2010 and 2013, the Ohio Department of Natural Resources recorded 190 complaints about water pollution. Of those, it confirmed six cases of pollution, although the ODNR argued that fracking was not the culprit in any of its confirmed cases.

  (Begos 2014)

- Between 2010 and 2014, West Virginia officials received 122 complaints about contaminated water wells. In four cases, the driller took corrective actions.

  (Begos 2014)

- Since 2004, officials at the Texas Railroad Commission have received more than 2,000 complaints about water quality related to nearby oil and natural gas operations. Texas regulators, however, have yet to confirm any examples of fracking related cross-contamination. (Begos 2014)

In each of the aforementioned studies and state records, attributing a causal connection between drilling, the subsequent fracking, and contamination is difficult because of a lack of pre-drilling record-keeping. It is also important to note that many complaints do not always distinguish between traditional wells or wells drilled since 2005, making the connection between contemporary fracking and water contamination more difficult.

Earthquakes and Production Wastewaters

Hydraulic fracking requires operators to fracture underground shale rock and tight oil sand formations, which causes thousands of micro-earthquakes. Most of these are never felt on the surface (USGS 2016, 2014). Seismicity-related concerns, although not directly connected to fracking, are related to a common method that operators use to dispose of their wastewaters, that is, injection wells (other methods include wastewater recycling and spreading the salty brines on roads [Davis and Fisk 2016]). In a small fraction of disposal wells (there are approximately 180,000 Class II underground injection wells), geologists have concluded that there is a complex causal link between fluid injection practices and the likelihood of causing a man-made quake (Ellsworth 2013; Kiger 2014). Essentially, the injected waters can exert enough pressure on known and unknown fault lines so that they slip, triggering an earthquake (Branstetter 2015; Bustillo and Gilbert 2015; Wilmoth 2014; Keranen et al. 2014; McGarr 2014; Oppel Jr. 2015).

Quality of Life

Industrial operations have the potential to disrupt residents' quality of life (Jacquet 2014; Kuster 2012; Ladd 2013). Rabe and Borick (2013) cataloged a number of possible ways that rapid oil and gas development may impact quality of life such as: increased truck traffic and overall vehicular congestion, strained infrastructure, increased levels of noise and dust contamination during construction and operations, and spikes in the number of spills. The Pennsylvania State Supreme Court, in its 4–2 decision that declared a portion of Act 13 unconstitutional, highlighted many of these 'quality of life' concerns. Before discovery of the Marcellus Shale, the majority opinion noted, many cities enacted zoning plans and other policies that reflected their citizens' preferences. Those policies created a reasonable expectation among residents as to what they may expect in terms of their quality of life, character of their community, and the environmental quality of surrounding areas. Act 13, according to the plurality opinion, "fundamentally disrupted those expectations, and ordered local government[s] to take measures to effect the new uses, irrespective of local concerns," *Robinson Twp., Washington County v. Commonwealth*, 83 A.3d 901 (Pa 2013).

Social science research is beginning to unpack the complex relationship between fracking's boom and bust economics and crime. Ruddell, Jayasundara, Mayzer, and Heitkamp (2014) observed that Montana and North Dakota counties impacted by development in the Bakken fields reported an increase in crimes compared with non-oil counties. Data from Archbold, Dahle, and Jordan's (2014) work also support a connection between rapid Bakken growth and an increased demand for police services.

Anecdotal evidence in North Dakota's Bakken Formation is also instructive. Diane Johnson, chief judge of the Mandan, Hidatsa, and Arikara (MHA) Nations, described crime as a tidal wave that has tripled since oil development in Western North Dakota began. She added that drugs are fueling much of the crime and are destroying the reservation (Horwitz 2014). Areas outside the reservation have also experienced a surge in criminal activity (drug and human trafficking), leading the FBI to open a new field office in Williston, North Dakota. Senator Heidi Heitkamp (D-North Dakota) in a statement noted that "Towns across the Bakken region have been hit hard by a spike in crime in the wake of our state's massive population boom . . . we need more robust federal support to fight lawlessness right in the eye of the storm—in Williston" (Heitkamp quoted by Cama 2015). Senator John Hoeven (R-North Dakota), added that "our new FBI [Federal Bureau of Investigation] office in Williston now has a site at the heart of the Bakken, which is vital not only for addressing crime, but also for maintaining our quality of life" (Hoeven quoted by Cama 2015).

Infrastructure

To extract oil and gas, operators need heavy trucks, equipment, and other industrial machinery, which can strain state and local transportation infrastructure. State transportation officials in New York, for example, estimated that if widespread production were permitted within the state, road repairs and improvements would cost the state between $211 and $378 million. The Texas Transportation Commission, in 2012, estimated that the roads above the Barnett and Eagle Ford formations already needed $40 million in repairs (Efstathiou Jr. 2012). Since 2009, Pennsylvania has collected more than $500 million in fees to pay for road inspections, maintenance, and repairs caused by the heavy trucks needed to transport millions of gallons of water, sand, and chemicals[1] (Salazar 2014).

The 'boom' may also lead to hastily designed and poorly planned growth (Jacquet 2014; Opsal and O'Connor Shelley 2014). During the bust times, overbuilt infrastructure quickly empties, becomes underutilized, and often neglected. Developers may also leave projects incomplete, which can exacerbate crime rates and levels of urban and suburban blight. Workers may also be left underemployed or unemployed and look to move on to the next 'boom' town (Jacquet 2014). Jim Weaver, a planner in Tioga County, Pennsylvania, described the boom-bust dynamic as "[dollars] dangling in front of you and you're chasing the carrot, before you know it you're out on a limb, and the limb gets sawed off" (Weaver quoted by DePillis 2015).

Employment Realities

The answers to questions such as how many permanent jobs can the industry support, how many indirect jobs will be created, and does the local population have the knowledge, skills, and abilities required by industry, are elusive and also subject to political debate. Proponents contend that oil and gas production creates thousands of jobs (Dlouhy 2011). Opponents counter that the employment statistics cited by supporters are inflated and that actual job growth is much lower and that most jobs are temporary (Songer 2014). For example, according to industry and its backers, the development in Ohio's Utica and Marcellus shale plays could have created more than 200,000 new jobs by 2015. A study by the Ohio State University, however, placed the actual job gains closer to 20,000 (Louis 2012).

Claims made during the 2014 Pennsylvania gubernatorial race illustrate how murky and challenging generating accurate employment data can be. The incumbent, Republican Tom Corbett, argued that unconventional gas extraction in the Marcellus supported around 200,000 jobs. The state's Department of Labor and Industry, however, reported only 30,000 additional natural gas–related jobs, a number cited by his Democratic opponent. Foran (2014) explained that both numbers could be true, noting that the differences between the two estimates are likely the result of two different definitions of what constitutes an industry-supported job. The Department of Labor's count likely included only those individuals directly involved in production (natural-gas extraction, well drilling, etc.). By comparison, the Corbett campaign's estimation was likely the result of a broader definition of job gains, which included all jobs created by natural gas's supply chain, including its downstream industries, and its impacts to the larger regional economy.

Well Cleanup

Oftentimes, a well's highly productive 'life' is over within a year (Jackson et al. 2014). The well's impact to the environment, however, can be much longer and related to whether it has been adequately capped and plugged. Three factors complicate the policy challenges related to well cleanup:

1. Many abandoned or orphaned wells were drilled long before the fracking-fueled boom and predate modern mapping technologies.
2. Many older (and less productive) wells are sold by the initial operator to small producers with fewer resources for cleanup.
3. The oil and gas renaissance has added to the state's inventory of wells that needs to be permanently closed.

The experiences of Wyoming and Louisiana are indicative of this dynamic. Wyoming is home to an estimated 1,200 abandoned oil and natural gas wells, and that number is expected to grow as production in the Jonah and Pinedale regions slows (Frosch 2013). According to state officials, the costs to plug existing wells could surpass \$10 million and that the responsible parties have never paid enough to cover those costs (Frosch and Gold 2015). Louisiana suffers from a similar well-plugging problem. Between 2008 and 2013, the state annually plugged around 95 wells but added approximately 170 to its list of abandoned/orphaned wells each year. Additionally, because of regulatory exemptions and other states policies, nearly 75% of wells drilled in Louisiana require no 'plugging' bond (Office of Conservation, Department of Natural Resources 2014). A state-by-state list of orphaned wells is available here: http://groundwork.iogcc.ok.gov/topics-index/orphaned-wells/state-progress

Land Fragmentation and Wildlife

Operators organize production into crisscrossing patterns that often transform large swaths of undeveloped land into smaller and disjointed parcels. The challenge of land fragmentation is particularly acute in the Eastern United States, which is home to environmentally sensitive forested areas (Pennsylvania and Ohio). The United States Geological Survey, in a recent study of energy development in Lycoming County, Pennsylvania, concluded that the area's undisturbed forests host several species that struggle near humans populations, provide insect control, purify water, offer residents and tourists recreational opportunities, serve an aesthetic role, and store carbon dioxide. Fracking-fueled land fragmentation, the report found, disrupts many of these core ecological services. The question of land fragmentation has also led to conflicts between environmentalists and industry. In Pennsylvania the oil and gas industry supported efforts to limit the power of state agencies to identify endangered species and that required industry to perform wildlife impact studies. Environmentalists support both positions (Cusick 2014; see also Slonecker et al. 2012).

Costs Summary

The environmental costs of fracking, according to critics, are pervasive and dangerous. Fracking pollutes air and water and exposes individuals to toxic chemicals. Even when environmental costs are minimal or can be managed, skeptics argue that benefits are overblown. They note that optimistic job, revenue, and greenhouse gas forecasts are misleading, temporary, and/or inaccurate. And, even if it brings new economic development to a state or local government, the boom and bust nature of resource extraction leaves communities vulnerable to increases in crime, blight, and unplanned growth.

Power and the Fracking-Fueled Oil and Gas Debate

The fracking debate is a compelling story. At the national and state level, fracking involves concerns and benefits that are similar across state lines. Yet, it also impacts each community differently. The fact that it simultaneously has national and local implications underscores the importance of identifying more optimal ways of sharing power and designing more effective governing arrangements to better manage the intersection of energy, economics, equity, and the environment.

Note

1 Pennsylvania uses these road-use pacts, also known as 'maintenance agreements', to maintain roads in areas with heavy truck traffic.

2 Oil and Gas at the Federal Level

How has federal involvement shaped the current shale renaissance and intergovernmental relations? How has the federal government impacted contemporary state-local relations? What powers do federal regulators hold when it comes to fracking and extraction? Should the federal government have greater or less oversight authority? What aspects of unconventional oil and gas production, transportation, and consumption do federal regulators manage and how does their authority interact with that of state and local governments? Has Congress acted? What executive agencies are involved and how does their involvement affect subnational units of government? If they are involved, what laws have lawmakers and regulators proposed in recent years that affect oil and natural gas production?

Federal legislators and regulators have entered the fracking debate. In 2005, Congressional lawmakers exempted unconventional oil and gas production from several federal environmental laws, enabling the states to vary significantly in how they balance fracking's costs and benefits and how they manage oil and gas production. Similar to supportive state and local officials, many Congressional Republicans cite energy security, job creation, adequate state oversight, and other economic benefits as reasons for their continued support (Songer 2014). They have authored legislation that supports state-centric regulatory roles, encourages additional production, and eases restrictions on exporting.

Opponents, typically House and Senate Democrats, often ground their opposition in the 'costs' included in Chapter 1. They have proposed laws to increase federal regulatory oversight, limit production in ecologically sensitive areas, and encourage more robust information disclosure (for wells on federal lands). They have also supported calls from opponents to mitigate environmental impacts such as limiting fugitive methane emissions and encouraging additional research. President Barack Obama, for example, has highlighted the myriad of environmental and public health risks and has pushed for additional federal oversight, especially related to production on federal lands. Despite supporting greater restrictions, the President's Clean Power Plan encourages the use of natural gas. The plan calls for "reducing the carbon intensity of electricity generation by shifting electricity generation from higher-emitting fossil fuel–fired steam power plants (generally coal-fired) to lower-emitting natural gas–fired power plants." Much of this natural gas, it should be noted, would not be economically recoverable without horizontal drilling and hydraulic fracturing (EPA 2016). Federal regulatory agencies such as the Environmental Protection Agency (EPA), the Federal Energy Regulatory Commission (FERC), and the Occupational Safety and Health Administration (OSHA) have also added to the multitude of 'voices' contributing to the oil and natural gas debate. Importantly, several states and local governments have pushed these agencies for additional oversight ranging from concern about oil tank cars on the nation's railroads to pipeline siting issues.

Following are several passages that reflect the varying attitudes and perspectives of federal officials:

- During his 2013 State of the Union address, President Obama stated that his "administration will keep cutting red tape and speeding up new oil and gas permits"

  (Obama 2013).

- Representative Rush Holt (D-New Jersey), in reference to the lack of state regulation regarding fracking wastewater pits, observed that federal regulation is needed to ensure that "the states have minimum standards."

  (Holt quoted by Kasperowicz 2013)

- Representative Bill Flores (R-Texas) defended the state-centric status quo by noting "states have been effectively regulating hydraulic fracturing on both government and privately owned lands for decades."

  (Flores quoted by Foran 2013)

- In Pennsylvania, Katie McGinty, the Democratic nominee in 2016 for the United States Senate, had described oil and gas production as "secret sauce" for job growth. One of her opponents in the primary supported a moratorium throughout Pennsylvania.

  (Yachnin 2016)

Federal involvement in oil and natural gas operations presents challenges and opportunities for state and local stakeholders. The legal exemptions from federal environmental laws have helped to carve out space for state-local fracking-related federalism to emerge. For state and local activists, the inclusion of federal agencies offers additional venues for activists to press for policy change if they are unhappy with existing subnational institutions. The federal government also commands greater resources available for research and other studies needed to inform state and local lawmakers and policy decisions. Yet, the federal government also presents opportunities for supporters to seek favorable legislation that protects the status quo. Therefore, it is imperative to examine the important federal roles and how they contribute to the ongoing intergovernmental debate related to fracking, drilling, and oil and gas production at large.

The Federal Government's Historical Role

Federal involvement in hydrocarbon regulation began in the early 20th century. In brief, several federal laws and state actions emerged during this era:

- The Public Holding Company Act of 1935 limited consolidation within the gas and electrical industries.
- Three years later, with the Natural Gas Act of 1938, federal lawmakers authorized the Federal Power Commission (FPC) (now the Federal Energy Regulatory Commission [FERC]) to establish "just and reasonable rates" for interstate sales of natural gas.
- In 1935, six producer states agreed to cooperate with one another to address common resource issues such as overproduction and waste. Congress ratified the agreement, known as the Interstate Oil and Gas Compact Commission (IOGCC). Today, membership in the IOGCC is composed of governors and/or their appointed representatives from producing states (all but a handful of New England and Upper Midwest states are at least associate members). Through the IOGCC, states coordinate their oil and gas production standards, identify conservation opportunities, research environmental remediation and plugging issues, and disseminate best management practices for industry (Interstate Oil and Gas Compact Commission 2015). The approach offered a number of potential advantages, as follows:
  1. First, it enabled states to avoid future federal rules.
  2. Second, by functioning as a 'block of states,' the IOGCC offered industry a greater degree of regulatory stability and predictability needed for long-term investment and planning.

  (Interstate Oil and Gas Compact Commission 2015)

Another wave of federal involvement in oil and gas production took place in the 1960s and 1970s, especially during the energy crises. Although this wave of federal involvement did not spark local defiance, it did nurture many of the technologies that have ushered in the shale renaissance and its attendant intergovernmental challenges:

- In the 1960s, under the Plowshare program, operators and the federal government began experimenting with how to fracture tight sand formations to 'rubbilize' and release oil and gas reserves with nuclear devices. In 1967, for example, a group of federal scientists, United States mining officials, and representatives of the El Paso Natural Gas Company 'rubbilized' deep, low-permeability, gas-holding formations underneath New Mexico via Gasbuggy, a 29-kiloton nuclear device.
(American Oil and Gas Historical Society N.d.)

- In 1974, President Ford signed the Energy Reorganization Act, which authorized the creation of the Energy Research and Development Administration (ERDA), the Nuclear Regulatory Commission (NRC), and the Energy Resources Council. Lawmakers charged the ERDA with promoting "unconventional gas research." The Department of Energy (DOE), ERDA's successor agency, continued to support research aimed at developing more efficient processes to extract unconventional fuels.
(Golden and Wiseman 2015)

- Under the Natural Gas Policy Act of 1978, lawmakers promulgated new and complicated rate-setting formulas designed to regulate the end-user price. In 1989, with the Natural Gas Wellhead Decontrol Act, lawmakers deregulated sales (including the sales of gas to a pipeline, local distribution company, end users, and as determined by FERC) of natural gas.
(Golden and Wiseman 2015)

Through the early 1990s, the industry benefited from federal tax incentives, public-private partnerships, and exemptions, several of which are outlined as follows:

- From 1980 to 1992, industry received the Section 29 Non-conventional Fuels Production Tax Credit. Golden and Wiseman (2015) estimated the credit's overall value at $10 billion.
- The DOE helped fund a number of demonstration projects in West Virginia (Appalachian Basin), Michigan (Antrim Shale), and Texas (Barnett Shale), targeting production from shale resources.
- Federal regulators assisted in the development of new technologies that are currently being used by oil and natural-gas operators such as three-dimensional micro-seismic imaging.
(Golden and Wiseman 2015)

Contemporary Federal Domestic Oil and Gas Legislation

The Bush and Obama administrations have contributed to intergovernmental conversations regarding domestic oil and gas production. Recent actions, such as the Energy Policy Act of 2005, have exempted industry from a variety of federal standards, as shown in Table 2.1. By removing itself from these portions of the fracking debate, the federal government ceded some of this 'space' to state and local governments. In some cases, this real or perceived lack of state oversight has contributed to local communities enacting ordinances that challenge, exceed, and/or defy comparable state regulations.

Table 2.1 The Federal Role in Fracking

Law	*Effect*
Federal Resource Conservation and Recovery Act (RCRA)	Federal law exempts many wells from being designated as superfund sites under RCRA. The law as written and interpreted by the EPA excludes petroleum (this includes crude oil), natural gas, and liquefied natural gas from being classified as a 'hazardous substance,' which is needed for superfund designation. Federal law also exempts "drilling fluids, produced water, and other wastes associated with the exploration, development, or production of crude oil or natural gas" from being designated as a hazardous substance (Brady and Crannell 2012; Soraghan 2015).
Clean Air Act (CAA)	Under the CAA, the EPA may aggregate smaller pollution sources, which are controlled by a single entity and located near one another. However, oil and gas wells are exempt from the aggregation requirement. Because most wells, by themselves, do not reach the CAA's threshold limit of hazardous emissions, operators are often not required to acquire a permit (Brady and Crannell 2012).
Clean Water Act (CWA)	The EPA, under its effluent guidelines program, prohibits oil and gas discharges to surface water and municipal sewer treatment plants.
National Environmental Policy Act	The Energy Policy Act exempted unconventional oil and gas operators from many requirements commonly included in environmental impact assessments. A second exemption, also enacted in the 2005 legislation, shifts the burden of proof to the public rather than the operators to prove that oil and gas activities are unsafe (Environmental Working Group [EWG] 2013).

Law	*Effect*
Safe Water Drinking Act	Under the 2005 Energy Policy Act, the Bush administration excluded operations from regulation and standards that are part of the Underground Injection Control program, unless diesel fuels are used (EPA 2010). Many states have enacted rules that are designed to minimize cross-contamination between drinking waters and industry.
Toxic Release Inventory	The EPA does not require that the oil and gas industry disclose chemicals to the Toxic Release Inventory (TRI). The EPA explains that much of the TRI data is already collected by applicable state agencies and that the agency does not have the capacity to collect site-specific information for the hundreds of thousands of wells that dot the U.S. landscape (FracFocus 2016).

Unconventional oil and gas operators do fall under Sections 311 and 312 of the 1986 Emergency Planning and Community Right to Know Act. The sections mandate that facilities that manufacture, process, or store designated chemicals produce a Material Safety Data Sheet (MSDS) for each chemical. The MSDS lists the properties and health effects of each chemical and must be available to state and local officials.

Since 2013, domestic oil and gas development has been a popular topic for members of Congress. As the sampling in Table 2.2 shows, both Democrats and Republicans have authored bills aimed at the industry, its economic and environmental impacts, and many of its intergovernmental actors. As the table demonstrates, some bills are clearly designed to impede or facilitate oil and gas production while others are designed to add additional regulatory requirements. Embedded within many of these bills is also language that has clear connections to the state-local disputes related to fracking. For example, under the proposed Protecting States' Rights to Promote American Energy Security, Congress would have mandated that the EPA objectively study fracking's impacts, including the evaluation of established risk management practices. These findings, in turn, would likely be used in the fracking debates taking place in states across the nation.

The oil and gas industry also benefits from several favorable provisions in the federal tax code. Here again, federal policy has encouraged production. As part of the Obama administration's FY 2015 budget proposal, the administration proposed to eliminate several incentives and subsidies available to industry. Three major incentives are noted in the following

Table 2.2 Proposed Fracking Legislation Since 2013

Number–Date	*Author*	*Name*	*Content/Summary*
H.R. 1902–2015	Jan Schakowsky (D-Illinois) Mark Pocan (D-Wisconsin)	Protect Our Public Lands Act	The bill would ban fracking on public lands.
H.R. 1154–2013	Jared Polis (D-Colorado)	Bringing Reductions to Energy's Airborne Toxic Health Effect (BREATHE) Act	The proposed bill would have changed the way permits are issued. Typically permits for air pollution are issued per well. The bill, had it passed, would have required an accounting of the 'aggregated' air quality impacts of drilling (i.e., air pollution from all their wells rather than a single well). The bill would have expanded the Clean Air Act's hazardous air pollutant list to include hydrogen sulfide.
S. 828 (2016)	James Inhofe (R-Oklahoma)	Fracturing Regulations are Effective in State Hands Act	Hydraulic fracturing on federal lands must comply with the law of the state in which the well is located. The law also names the states as the "sole authority to promulgate or enforce any regulation, guidance, or permit requirement regarding hydraulic fracturing on or under any land within their boundaries" (Congress 2016). A similar bill was also introduced by Senator Inhofe in 2012.
H.R. 1175–2013	Matt Cartwright (D-Pennsylvania)	Focused Reduction of Effluence and Stormwater Runoff Through Hydraulic Environmental Regulation (FRESHER) Act	The proposed bill would have required that oil and natural gas companies acquire stormwater runoff permits. It would have also included resources for a study on fracking's impact to surface water.

Number–Date	*Author*	*Name*	*Content/Summary*
S. 735–2015	Sen. Robert Casey, Jr. (D-Pennsylvania) Reps. Diana DeGette (D-Colorado) Chris Gibson (R-New York)	Fracturing Responsibility and Awareness of Chemicals	Under the proposal, natural gas firms, before beginning operations, would be required to acquire permits under the Safe Drinking Water Act. A second major piece of the law would force drillers to disclose to the public the chemicals used to frack each well.
H.R. 2728–2013	Bill Flores (R-Texas) Henry Cuellar (D-Texas)	Protecting States' Rights to Promote American Energy Security Act	The proposed bill would restrict the Department of Interior from enforcing federal fracking rules in states that already exercise regulatory and oversight authority relative to fracking. The bill would also require that the EPA "objectively" study fracking's impacts; the study must include evaluating established risk management practices of industry and state regulation.
H.R. 1900–2013	Mike Pompeo (R-Kansas)	Natural Gas Pipeline Permitting Reform Act	The proposal would mandate that the FERC make a decision (to approve or reject) about proposed natural gas pipelines within 12 months.
H.R. 1965–2013	Doug Lamborn (R-Colorado)	Federal Lands Jobs and Energy Security Act	The proposed bill included language that shortened the deadlines for federal agencies to approve drilling permits on federal lands. The bill would have also established a minimum on the total number of leasable federal land for oil and gas development. A third major provision would have imposed costly fees on protesters who lodge formal complaints/appeals relative to permits.

(Continued)

Table 2.2 (Continued)

Number–Date	*Author*	*Name*	*Content/Summary*
H.R. 4155–2013 H.R. 6–2013	Ted Poe (R-Texas) Cory Gardner (R-Colorado)	Fight Russian Energy Exploitation (FREE) Act (Poe) Domestic Prosperity and Global Freedom Act (Gardner and others)	The two bills are interrelated. Poe's bill, the Fight Russian Energy Exploitation (FREE) Act, adds several Central Asian nations, the European Union, and the Ukraine to the list of countries eligible for expedited approval in order to receive liquefied natural gas exports (LNG). Gardner's proposal adds World Trade Organization (WTO) nations to the expedited LNG export list. It also would trigger the immediate approval for all pending applications (which have been listed in the Federal Register).

Sources: Berwyn (2013), Cama (2014), Congress (2015, 2014, 2014a), Kasperowicz (2013)

list, and the complete list is available here: www.treasury.gov/open/Documents/USA%20FFSR%20progress%20report%20to%20G20%20 2014%20Final.pdf (US Treasury 2016).

- Intangible Drilling Costs—Oil and gas companies may deduct intangible drilling costs, which include specific drilling and development costs for domestic oil and gas wells.
  (Annual cost $1.495 billion)

- Depletion (Percentage and Cost)—According to the IRS, depletion is the consumption of natural resources "by mining, quarrying, drilling, or felling." Once those minerals are extracted, the IRS permits the landowner to make an income tax deduction.
  (Annual cost $1.343 billion)

- The Manufacturing Deduction—This subsidy is available to taxpayers who engage in specific production activities, including the production of fossil fuels. The current deduction "for income from oil and natural gas production activities" is set at 6 percent.
  (Annual cost $1.250 billion)

Fracking and Federal Rulemakers

Federal regulatory involvement extends to exports, rail safety, worker health and well-being, pipeline siting and locations, and drilling requirements on federal lands (Mufson 2013). For some of these issues, state and local governments have called for and supported additional layers of federal regulation. Yet, in other areas, state officials have described federal actions as an overreach and as unnecessary. Dissatisfied states have also sued the federal government in the United States District Court of Wyoming for overstepping state legal authority (regulation on public lands). Despite the threat of lawsuits, in a variety of issues related to fracking, federal rulemaking continues, and is outlined in the following sections.

Fracking and the Environmental Protection Agency

The EPA's role in the fracking debate includes both research and oversight. It has performed and released numerous studies about impacts on drinking and surface water resources, air quality, and methane emissions. Activists on both sides of the fracking debate have seized these findings to support their arguments. Such debates, it should be noted, have taken place in both statehouses and city halls.

The EPA has also promulgated rules that impact the industry:

- Under the 2016 rule, operators must capture any escaped methane emissions. Based on EPA estimates, the rule will reduce industry's methane footprint by approximately 510,000 short tons and lead to $690 million in savings from waste reductions in 2025. The EPA anticipates that industry costs will total approximately $530 million in 2025. The EPA has also issued an Information Collection Request (ICR), soliciting information and projections from industry that includes equipment and emissions control installation costs as they relate to existing wells.
  (Davenport 2016)

- Under its Underground Injection Control (UIC) program, the EPA oversees Class II disposal wells that are often used to dispose of waste fluids generated as part of oil and gas production.
- EPA regulations now prohibit "discharges of wastewater pollutants from onshore unconventional oil and gas (UOG) extraction facilities to publicly owned treatment works."
  (EPA 2016)

Fracking and the Railroads

The prodigious rise in domestic oil and gas production has impacted all modes of transportation, but perhaps none more so than the nation's

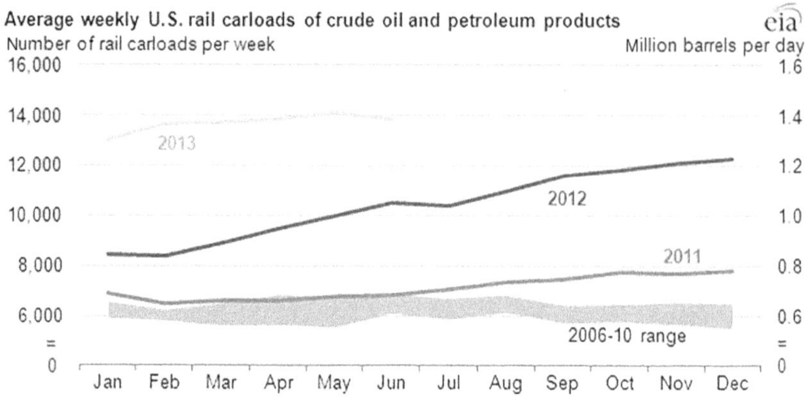

Figure 2.1 Average Railcar Loads

Note: Petroleum product rail shipments do not include ethanol. Conversion of rail carloads per week into million barrels per day includes assumption of 700 barrels per rail carload.

Source: U.S. Energy Information Administration, based on the Association of American Railroads.
EIA (2013b)

railroads (Lavelle 2015).[1] In 2013, crude oil shipped via the railroads had grown by 255 percent compared with the early 2000s, which has precipitated local-level concerns (National Conference of State Legislatures 2014). This growth is reflected in Figure 2.1, which shows the average number of railcars used for shipping oil and gas by year.

Oil trains, often originating from the Bakken fields in North Dakota, are increasingly rumbling through major population centers of Chicago, Philadelphia, and Albany (Mouawad 2015). Major routes are shown in Figures 2.2 and 2.3. The increase in production, as measured by the 'thickness' of the gray line suggests that thousands of more crude oil barrels are crisscrossing the country in 2014 compared with 2010, meaning that many more trains are passing through urban centers.

Historically, the mechanics of crude oil transportation have rarely made the front page of any newspaper. Yet, since 2013, a small but growing number of high-profile derailments have heightened state and local interest in and attention to rail safety. Table 2.3 lists four recent derailments along with the scope of the disaster, including (when available) the number of cars, gallons spilled, and injuries (Fitzpatrick and Griffin 2015).

State and local politicians have pressed for greater railcar safety (Phillips 2015). In response to the increasing number of oil trains rumbling through Chicago suburb Barrington, Illinois, Village President Karen Darch suggested that it would be helpful if industry shared more information about the number, location, and frequency of oil cars with

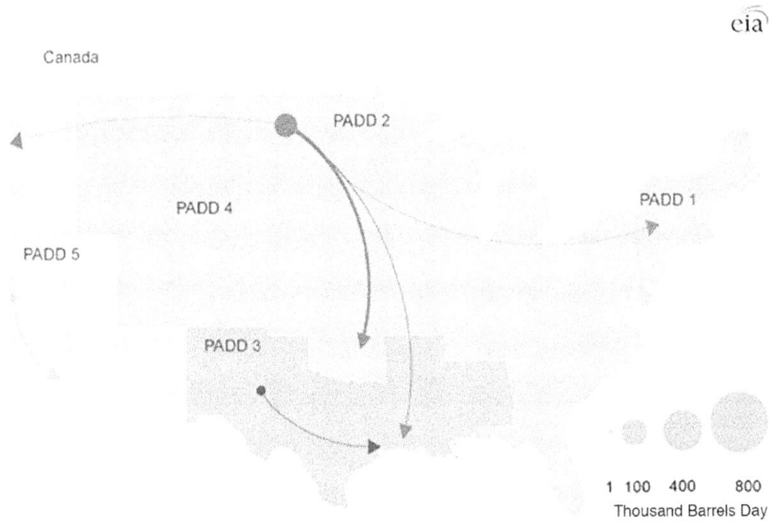

Figure 2.2 Crude by Rail Movements, 2010
PADD, Petroleum Administration for Defense District
Source: EIA (2015b)

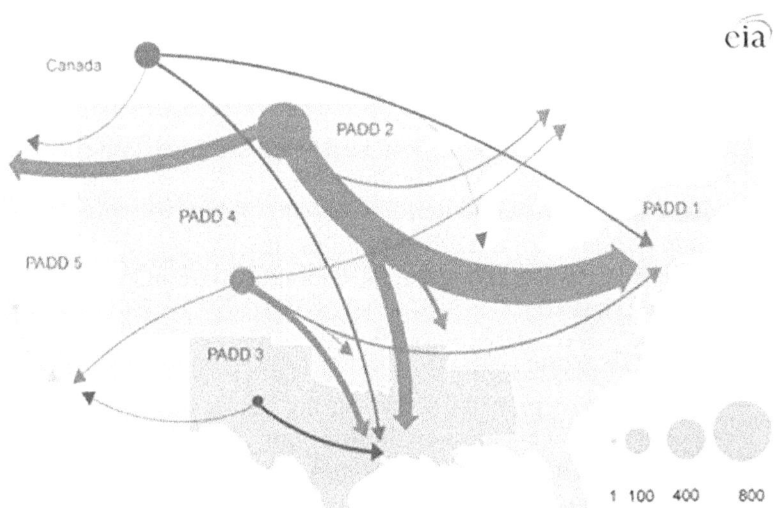

Figure 2.3 Crude by Rail Movements, 2014
PADD, Petroleum Administration for Defense District
Source: EIA (2015b)

Table 2.3 Crude Oil Shipping and Rail Safety

Location	*Date*	*Number of Cars*	*Gallons Spilled (in millions)*	*Notable Impacts*
Lynchburg, Virginia	April 2014	17 of 105 oil tank cars	Data unavailable	300 residents were evacuated
Lac-Megantic, Quebec	July 2013	63 of 72 oil tank cars	1.5	47 deaths
Aliceville, Alabama	November 2013	26 of 88 oil tank cars	.75	Data unavailable
Casselton, North Dakota	December 2013	20 of 106 oil tank cars	.4	Data unavailable

Source: Fitzpatrick and Griffin (2015)

local officials, first responders, and citizens. Tom Weisner, the mayor of Aurora, Illinois, also shared concerns about the risks posed by oil railcars: "I would say our chances of having some kind of issue [relating to rail incidents] in Aurora have gotten worse by about 4000 percent" (Weisner as quoted by Wronski 2014).

State and local concerns have trickled their way up to administrators at the United States Department of Transportation. Despite industry protests, in 2015, the Department of Transportation, through a collaboration with the Pipeline and Hazardous Materials Safety Administration (PHMSA) and the Federal Railroad Administration (FRA) announced new rules addressing the following (DOT 2015, 2015a) (see also Mouawad 2015; Tita 2015):

- Established a new tank car standard based on safety and risk.
- Established a risk-based retrofitting schedule for older tank cars carrying crude oil; retrofitting for crude oil tank cars must be complete within two years of the rule going into effect.
- Established a new braking standard for certain trains; includes trains carrying crude oil.
- Established new safety protocols, such as route restrictions and speed limits, for trains carrying large volumes of flammable liquids.
- Established new procedures for sharing information with local first responders. Railcars must establish a point person for interested parties (including local governments) to facilitate more efficient sharing of information.
- Established a new set of safety standards for trains considered "high-hazard flammable," which are defined as "a continuous block of 20 or more tank cars loaded with flammable liquid or 35 or more tank cars loaded with a flammable liquid dispersed through a train." Under the new rules, these trains must travel at slower speeds.

(DOT 2015, 2015a)

Fracking and the Occupational Safety and Health Administration

Even before the recent oil and natural gas boom, Congress had empowered the Occupational Safety and Health Administration (OSHA) to protect the industry's workers. Under 29 Code of Federal Regulation (CFR) 1910, OSHA may issue rules that mitigate the hazards present during site preparation, construction, trenching, excavation, drilling, well servicing, and storage (OSHA's construction standard, 29 CFR 1926) (OSHA 2015, 2015a). OSHA standard 1910.95a-b also mandates that operators protect workers from being exposed to excessively loud sounds/volumes during production (Colorado Department of Public Health and Environment [CDPHE] N.d.).

Although not necessarily a direct cause for intergovernmental conflict, oil and gas development in urban and suburban communities has contributed to a relatively high number of accidents involving both civilians and industry workers (Jervis 2014). Between 2003 and 2010, approximately 823 oil and gas workers died in on-the-job accidents. This number is significantly higher than many other occupations but still a tiny fraction of all oil and gas employees (there are several hundred thousand oil and gas workers in any given year). Specific threats to workers' safety include: vehicle accidents; being struck by, caught in, or caught between objects/equipment; explosions and fires; falls and slips; being trapped in a confined space; and exposure to temperature extremes (Banerjee 2014). The relative frequencies, as a percentage of overall industry injuries, are shown in Table 2.4:

Table 2.4 Frack Sites and Injuries

	Death	*Injury (Colorado) 2012 and 2013*	*Injury (national) in 2010*
Most frequent	Highway crashes (29.3%)	Strains (35.3%)	Being struck by/against an object (35.8%)
Second most frequent	Being struck by an object (20.1%)	Slips, trips, or falls (24.1%)	Caught in an object, equipment, material (21.2%)
Third most frequent		Struck or injured by an object (14.1%)	Falls on same/lower level (14.6%)
Fourth most frequent			Overexertion (11.5%)
Fifth most frequent			Exposure to harmful substance (3.3%)
Sixth most frequent			Slips or trips (1.3%)

Source: Colorado Department of Public Health and Environment [CDPHE] (N.d.)

OSHA also oversees worker safety during frack sand operations such as using sand movers, transfer belts, and blenders. There is also risk when workers move the sand to the well site. The National Institute for Occupational Safety and Health (NIOSH) detected the highest levels of silica dust exposure for workers involved in sand moving and blending. Inhaling silica puts workers at risk for developing silicosis, scarring, and difficulty breathing.

Fracking and the Pipelines

Three federal agencies regulate the nation's pipeline inventory. Congress housed the Pipeline and Hazardous Materials Safety Administration (PHMSA) within the Department of Transportation and ordered it to oversee pipeline safety and development (PHMSA 2014). Lawmakers charged the Federal Energy Regulatory Commission (FERC) with setting distribution rates and addressing siting issues (interstate pipelines, compressor stations [needed to push the gas through the pipeline], and storage facilities). Finally, lawmakers authorized the Transportation Security Administration to enact policies to ensure pipeline security. A side-by-side comparison of the three agencies is presented in Table 2.5:

Table 2.5 Comparing the Federal Regulatory Agencies

Pipeline and Hazardous Materials Safety Administration	*Federal Energy Regulatory Commission*	*Transportation Security Administration*
Pipeline safety rules	Rate-setting methods	Pipeline security
Pipeline development and expansion	Rules for business practices	
Certification that a pipeline is safe before operations can begin	Siting and construction	
	Operations of pipelines (compressors)	
	Natural gas storage	
	Liquefied natural gas (LNG) facilities	
	Lead agency during the reviews required by applicable environmental laws such as the National Environmental Policy Act, the Endangered Species Act, and the National Historic Preservation Act	

Source: EIA (2015a), Interstate Natural Gas Association of America (2013), Parfomak (2013)

The distribution network begins with flow and gathering lines, which transport oil and gas from the field to storage facilities and then onto a processing facility, which removes impurities and separates the fluid into its various components. Once cleaned and separated, the fluid then enters the larger constellation of interstate lines. This network:

- Consists of 300,000 miles of interstate transmission pipelines
- Transported more than 36 trillion cubic feet (Tcf) of natural gas
- Comprises interstate lines that are approximately 71 percent of all transmission lines in the United States
- Is chiefly utilized by the 30 largest pipeline firms, which own three quarters of all interstate pipeline mileage and approximately 72 percent of the total distribution capacity (183 billion cubic feet)

(EIA 2008)

- Includes two million miles of smaller, low-pressure distribution pipelines that crisscross American cities and towns to deliver gas to residential, commercial, and industrial consumers (Union of Concerned Scientists 2015). These smaller lines are often regulated at the state and local levels

(Seville 2014)

A map is available at the following URL: www.eia.gov/pub/oil_gas/natural_gas/analysis_publications/ngpipeline/interstate.html

Pipeline siting and management has also sparked episodes of local pushback, as well as questions over safe but economically efficient setback distances. In 2016, for example, the Massachusetts Senate advanced legislation that established a 1,000-foot setback between new pipelines and schools, senior centers, and residential neighborhoods (Dumcius 2016).

Fracking and the Bureau of Land Management

Lawmakers authorized the Bureau of Land Management (BLM) to manage oil and gas production on federal lands. Besides overseeing sales, BLM staff conducts preliminary environmental analyses, enacts rules, and enforces applicable regulatory provisions. In 2016, the agency oversaw approximately 63,000 onshore wells (BLM 2016; EIA 2015). Industry has remitted millions in lease payments for extraction on BLM-controlled public lands.

In early 2015, the Obama administration announced new rules governing the fracturing of wells located on federal lands. The rules, which are being challenged by several states, would require (Neuman 2015):

- Operators to submit information about the chemicals used during the fracturing process to FracFocus, an industry-supported website, within 30 days of completing operations. Additionally, they must also provide the Interior Department with data relative to the geology of lands adjacent to the well.

- Federal employees and/or contractors to inspect the concrete barriers that separate the well from surrounding groundwater and other underground formations.
- Operators to comply with new safety standards for chemical storage.

Wyoming (later joined by North Dakota, Colorado, and Utah) have filed suit against the BLM seeking to block the rule's implementation (Cama 2015a). Wyoming Governor Matt Mead, a Republican, explained that the lawsuit protected his state's regulatory authority and that "states like Wyoming should be rewarded for their leadership, not be punished by having additional layers of regulation" (Mead as quoted by Storrow 2015). In September 2015, the United States District Court in Wyoming issued a preliminary injunction (later made permanent in a June 2016 decision), halting the BLM's implementation of the rules (Finley 2015). (CRS Reports & Analysis. 2015).

Western Native American Tribal leaders have also criticized the BLM's approach. Southern Ute Business Committee Chairman Shaun Chapoose stated that the rules "infringe upon tribal sovereignty." He added that the BLM's actions limited his tribe's economic opportunities and impaired its ability to practice self-government. Other Ute leaders have criticized the BLM's decision to apply the rule uniformly, rather than developing rules on a case-by-case basis (Liesik 2015).

The oil and natural gas boom has also arrived in the federal forests. Federal well sites located on Forest Service lands are governed in 'two pieces.' Subsurface requirements must meet applicable BLM regulations. The United States Forest Service monitors and enforces surface operations through a Surface Use Plan of Operation (SUPO). Oil and gas politics in the nation's forests have become increasingly contested, as noted in the following examples:

- In 2012, the BLM announced plans to lease 43,000 acres of the Talladega National Forest in Alabama for natural gas exploration. Following public opposition, the BLM rescinded the offer, stating that it needed more time to collect public input before issuing any drilling permits.

  (Gang 2013)

- In 2012, the Wayne National Forest in Ohio reported approximately 493 active and producing wells (mostly traditional wells) (United States Forest Service 2015). In 2016, the federal government announced that it planned to auction additional acres.

- In December 2013, the BLM announced plans to auction approximately 200 acres in the Homochitto National Forest in Southwestern Mississippi for oil and gas exploration.

  (Gang 2013)

- In 2013–2014, George Washington National Forest planners stated that the forest's updated management plan would include drilling. Following public protests, forest managers agreed to ban fracking in approximately 90 percent of the forest, with drilling allowed on 177,000 acres, the vast majority located atop of privately held mineral estates.

  (Gabriel 2014)

Summary

The federal government's relationship to oil and natural gas dates back to the early 20th century. Despite the regulatory focus on price controls ending in the 1980s, federal rulemakers still play a large role in shale oil and gas politics. Somewhat quietly, the federal government has nurtured many of the technologies that set the stage for the boom and has invested billions of dollars through incentives, research, investments, and tax breaks to aid in the development of unconventional oil and gas resources. Its regulatory involvement, remains largely focused on federal lands, pipeline security, worker safety, methane emissions, and rail traffic. Importantly, the federal government has yet to become heavily involved in many of the issues at the core of state-local disputes (the subject of following chapters).

Note

1 The Federal Railroad Administration has jurisdiction over track safety, rail crossings, railroad equipment, and other operating procedures. The agency collaborates with the Pipeline and Hazardous Materials Safety Administration (PHMSA), the Department of Homeland Security's Transportation Security Administration (TSA), and the National Transportation Safety Board (NTSB) to manage the transportation of crude oil via rail (National Conference of State Legislatures 2014a).

3 Oil and Gas at the Subnational Level

What are the contours and parameters of subnational legal relationships? Is there variation across the states? How are state and federal environmental and energy policies implemented? What are the roles of state and local governments when it comes to the environment and oil and gas production? What aspects of unconventional oil and gas development have mobilized local leaders and stakeholders to engage in acts of defiance? How widespread are state-local controversies relative to fracking and hydrocarbon production? What options are available to local governments should they seek to limit oil and gas development or fracking as an extractive technology? What are the risks and rewards to local leaders or communities should they choose to enter into the fracking fracas? The answers to these questions are critical to understanding the power struggles behind the oil and gas disputes and to identifying which level of governance is best positioned to respond.

The Intergovernmental Fracking Puzzle

Identifying the optimal locus of responsibility is a difficult question in any policy domain. Oil and gas production is no different (Davis 2012; Fisk 2016). In general, supporters of oil and gas development have argued for a state approach. The Colorado Oil and Gas Association's (COGA) position on well setbacks is typical of the logic behind arguments for state control: "[It] is not possible to safely and efficiently develop the state's mineral resources under a patchwork of inconsistent local rules" as oil and gas deposits do not follow city line (Schuller as quoted by COGA 2012). COGA has also argued that local governments lack the technical expertise needed for effective regulation and oversight (Phillips 2012). Local officials have responded by noting that they know their citizens' preferences best and because they feel the impact of development most directly, they are ideally 'located' to contribute to discussions on extraction-related governance. Local officials also point to their traditional role of regulating land use and other industrial activities as additional justification for a regulatory role. In effect, both levels of government have powerful claims to be the primary regulator of oil and natural gas development. Because of this, each must be involved in identifying and implementing new solutions to the challenges associated with oil and gas during the fracking renaissance. When state and local leaders fail to cooperate and work together, conflict is likely.

The State-Local Legal Lines

Unlike federal-state intergovernmental issues, state and substate government relationships exist in a policymaking environment that is created, governed, and policed by state leaders. For the most part, states have the power to determine the legal boundaries of local discretion, accountability, and autonomy (see Mashaw 2006 for a discussion of the differing forms of accountability). They also establish the scope of municipal responsibilities and when city governments may act without prior state authorization (Bowman and Kearney 2011). Additionally, states may limit the policy options available to local leaders or they may require that local governments act in very specific ways. Finally, they may also use their legal authority to induce a desired policy or to sanction cities when they fail to meet a standard set by state lawmakers (Cremer and Palfrey 2002).

Despite a comparative lack of formal power, cities and counties are not without political and legal options to effectuate policy change (Berman 2003; Zimmerman 1995, 2012). Local governments pass ordinances and author land-use plans designed to protect public health and safety. Riverstone-Newell (2012) also points out that municipal governments can bargain and negotiate with each other (to gain leverage and form alliances) and state lawmakers. Additionally they may directly lobby state lawmakers and regulators, especially during administrative rulemaking processes. If 'insider' tactics fail, community leaders may participate in protest movements, gather data/information and disclose it publicly, write press releases, pass resolutions, and use social/traditional media to pressure state leaders (Berman 2003; Sherman 2011; Zimmerman 1995). Finally, local governments can also use more passive means to influence state policy by strengthening their reputation/expertise, improving their organizational capacities to address oil and gas issues, and partnering with other communities to develop regional-wide approaches to addressing production.

Dillon's Rule and the Cooley Doctrine

The Dillon's Rule and Cooley Doctrine are often considered as foundational for understanding the allocation of state and local powers and how courts adjudicate state-local disputes (Wright 1978). Under the Dillon's framework, the state is considered supreme and political power begins and ends in the statehouse (Bowman and Kearney 2011; Hodos 2009).

The Cooley Doctrine presumes that local governments hold inherent self-determination powers (Berman 2003; Bowman and Kearney 2011; Csoka 2007; Krueger and Bernick 2012).

In 1868, Judge Dillon first articulated the theory of state supremacy when he famously declared: "Municipal corporations owe their origin to, and derive their powers and rights wholly from, the legislature. It breathes into them the breath of life, without which they cannot exist . . . as it creates, so it may destroy" (*City of Clinton v. Cedar Rapids and Missouri River Rail Road Company*). The United States Supreme Court restated much of this logic in the 1907 case, *Hunter v. City of Pittsburgh*, 207 U.S. 161. This state-centric approach has multiple implications for intergovernmental relationships, as shown in Table 3.1.

Under Cooley, local governments possess "home rule" powers, meaning that they enjoy the right of self-government, including home rule

Table 3.1 Impacts of Dillon's Rule

Institution	*Area*	*Example*
State courts	Defining rules of construction	Courts use Dillon's Rule as a way to interpret cases that involve intergovernmental (state to local) grants of authority by presuming that local authority is limited to those areas previously identified by state laws and regulations. This could lead to court decisions that restrict local authority in various areas such as the establishment of setbacks, site security measures, and venting/flaring restrictions if they are not clearly authorized to act by the state.
Statehouse/local	Heightened the intergovernmental lobbying burden	City officials must expend their time and other resources to lobby state lawmakers to approve and reject bills. The National League of Cities (2013) estimated that in an average year in a Dillon state—local governments file approximately 2,000 special acts, requests, and exemptions from state leaders. This could require local government officials to lobby state lawmakers as unique issues related to development arise.
Local	Modify city authority to regulate and act in policy areas	State officials could narrow or enlarge city authority in a variety of issue areas related to oil and gas development; these include aesthetics, transportation, security, public safety, etc.
Local	Discouraging local innovation	Richardson Jr. (2011) described that local governments may be hesitant to innovate and are likely to be overly cautious about using their full authority to tackle public issues. This may lead to conflict as local communities cannot respond to citizen demands.

Source: Bowman and Kearney (2011); Author generated

authority and the right to be left alone by state lawmakers (Berman 2003; National League of Cities 2013; Richardson Jr. 2011; see also Welch 2007). The National League of Cities (NLC) has identified four basic types of home rule authority:

- Functional home rule involves the authority to exercise local self-determination/government (i.e., home rule as applied to specific local-only policy arenas).
- Fiscal home rule involves the power to identify and utilize revenue sources, establish tax rates, issue and repay debt, and other financial matters.
- Personnel home rule involves the ability to promulgate and enforce employment rules, including collective bargaining practices.
- Structural home rule authority involves questions over the structure of local government. In other words, if a local government enjoys structural home rule authority, it may select its own form of government and the language of its charter, and it may revise its charter.

The Cooley Doctrine also impacts state and local operations in a number of ways, as shown in Table 3.2.

Table 3.2 Impacts of the Cooley Doctrine

Institution	*Area*	*Example*
State courts	Increasing intergovernmental litigation	State and local actors are often uncertain as to what constitutes a 'state' issue, what is a 'local' issue, and what is a joint 'state-local' issue. If issues such as oil and natural gas extraction have state-local mixed impacts, is there clarifying language that prescribes the extent of state/local authority and if there is material conflict between the state and local laws (Berman 2003).
Local governments	Depending on language used, Cooley states may encourage local innovation	Despite decentralization, municipal powers can still be restricted through state action but since local governments can act without prior authorization they may be able to respond to challenges as they arise (Berman 2003).
State and local governments	Compounds difficulties with coordination	From questions of air pollution to local economic development practices, many local actions have extraterritorial impacts (i.e., municipal decisions impact other communities). It has been difficult for local leaders to coordinate these 'regional' impacts and to identify state/local boundaries, especially without state encouragement (Richardson Jr. 2011).

Source: Author generated

Richardson Jr. (2011) identified three additional policy-specific dimensions that shape state-local legal relationships: location/region, issue area, or city characteristics. A local government, for example, may be authorized to enact economic development incentives and utilize zoning policies but lack the authority to rely on specific tax incentives/rebates without state authorization. States may also permit larger local government greater degrees of functional home rule authority (Boscarino 2013; Bowman and Kearney 2012; Richardson Jr. 2011). The trans-jurisdictional geology of many natural resource deposits adds an additional layer complicating the intergovernmental management of oil and gas production, as shown in Table 3.3.

Table 3.3 The Complexity of Intergovernmental Relationships

		State Differences	*Municipal Differences*
First layer	Interstate	Variations among two or more states, assuming the issue cuts across multiple state-lines	NA
Second layer	Intrastate	Variations between **broad** policy areas such as land use, fiscal, public safety, etc.	Variations between or among cities/counties; cities and other cities
Third layer	Intrastate	Variations **within** policy areas such as the ability to issue comprehensive zoning maps but lack the authority to determine if a variance is necessary	Variations between or among cities/counties

Source: Author generated

The Intergovernmental Landscape

The cause of and the solutions to local defiance are shaped by broader trends in subnational environmental regulation, place-specific factors, and variables unique to oil and gas development. City halls are implementing programs aimed at urban sustainability and environmental protection. They have worked to attract green energy jobs and have encouraged the use of renewable energy. These priorities and related values may then be at odds with opportunities to support oil and gas development. Local communities also have 'other' programmatic responsibilities that shape the scale, scope, and pace of environmental outcomes and commitments such as brownfield remediation, waste management and recycling, drinking water management, and energy delivery (Abel, Stephan, and Kraft 2007; Agranoff and McGuire 2001; Betsill 2001; Betsill and Rabe 2009; Daley and Garand 2005; Daley and Layton 2004; Davis 2014, 2012;

Duroy 2008; Kraft, Stephan, and Abel 2011; Klyza and Sousa 2007; Matisoff 2008; McCright and Dunlap 2011; Ringquist and Garand 1999; Scheberle 2004). They also have unique experiences with environmental management and vary in the degree to which they are inclined to engage in environmental protection efforts.

There are issue-specific reasons that may underscore local skepticism and pushback towards oil and gas development. Environmental or public health concerns resulting from wells located near residential neighborhoods, schools, and other occupied structures may push some communities to enlarge setback distances. Municipal leaders may also have concerns over finance-related issues. This may include impacts to residential property values but also, to the extent to which they receive resources, to offset the costs of development. Finally, other sources of local defiance may relate to emergency management and drinking water safety, and how information is shared among state, local, and industry personnel.

State and Local Environmental Policymaking

Beginning in the 1980s, federal and state policymakers began experimenting with alternative approaches to environmental management (Davis 2005; Scheberle 2004; Scholz and Wang 2006). Reforms stressed flatter organizational hierarchies and incorporated the expertise of private actors and subnational governments (Hempel 2009; Lehner 1993; Ringquist 1993; Vig and Kraft 2010). Environmental policymaking in 2016 reflects many of these changes and involves multiple levels of government, a variety of private and public actors, and a web of incentive programs and alternatives to traditional command and control regulation (Woods and Potoski 2010). Although the incorporation of additional actors may increase overall transaction costs, it is also a necessary ingredient for effective regulation, collaboration, and intergovernmental cooperation (Konisky and Woods 2012; Rabe 2006; Scheberle 2004; Woods and Potoski 2010).

In addition to their roles as partners in the implementation of many federal environmental laws, state and local governments also deliver a variety of services that have significant environmental impacts. State and city governments enact land use management plans, issue laws/ordinances, pass building codes, deliver electricity, and write zoning laws (Portney and Berry 2010). They also build roads, bike lanes, parks, trails, natural spaces, and other facilities. Each of these activities shapes the location and intensity of economic development and the degree to which the environment is protected (Betstill 2001; Trisolini 2010). These responsibilities likely have a number of impacts that may shape state-local relationships:

1. As local government professionals perform additional services, they grow in their competencies and abilities to perform such services, which may make the local government more autonomous and independent and less dependent on the state.
2. As local governments perform environmental services, they may outpace or disagree with state-level priorities. Because oil and gas impacts are felt differently across jurisdictions, local skeptics may come to believe that a "state average" cannot adequately mitigate the impact of oil and gas "on the quality, quantity, and well-being of our natural resources" on their particular community, *Robinson Twp., Washington County v. Commonwealth*, 83 A.3d 901 (Pa 2013).
3. State and local officials interact with one another (Scheberle 2004). They collect, share, and receive information that impacts one another although the degree of salience attached to those pieces of information may vary (Berry 1989; Browne 1996; Heclo 1978; Krueger and Bernick 2012; O'Toole 1997). Pressman and Wildavsky (1984) add that implementation conflicts are likely to emerge when networks of public agencies, private providers, and non-profits are involved (Peters 2011; Sherman 2011).

Some local-level governments have also committed themselves to priorities that balance economic, environmental, and social equity goals, which they view as at odds with oil and gas development (Dryzek 2005; Feiock and Kim 2001; Feiock and Stream 2001; Hempel 2009; Kamienicki 2006; Keller 2009; Layzer 2002; Vig and Kraft 2009). For many of these communities, the practice and pursuit of urban sustainability has led to a variety of learning opportunities for citizens, voluntary and regulatory programs, and incentives and sanctions to encourage or mandate change (Adams 2006; Kates, Parris, and Leiserowitz 2005; Opp and Saunders 2013; Portney 2003; Saha and Paterson 2008; Zarsky 2010). The commitment to sustainability may provide additional opportunities for those skeptical of oil and gas production to organize and exchange information, and to press local leaders for more restrictive policies.

Citizen Engagement and Access

Local governments are generally more accessible to citizens. Because of this, citizens have more opportunities to share their preferences with lawmakers and to push for policies reflective of those preferences. In some policy arenas, this has contributed to local governments adopting policy positions that run counter to their state government. Portney (2009) posited that council members and city personnel are particularly well positioned to respond to citizens' concerns as they relate to sustainability. By listening and acting upon local preferences, city officials may intentionally or even unintentionally create an oppositional relationship with state leaders (i.e., an alignment between city and state policies no longer exists). Access to ballot measures also impacts state-local relationships. Many states and localities provide citizens with opportunities to bypass their representatives in favor of ballot initiatives. In fact, ballot initiatives led to the cities of Fort Collins and Longmont, Colorado, and Denton, Texas enacting policies that the industry argued conflicted with state policy (Fisk 2016).

Differing Priorities

Intergovernmental conflicts may also result if city leaders hold differing priorities compared with state leadership. Such a scenario underscored the battles between the city of Houston, Texas, and state policymakers (Bruhl, Linder, and Sexton 2013). Examining the city's efforts to improve its air quality, Bruhl, Linder, and Sexton (2013) found that the intensity of intergovernmental conflict was related to the type of action/strategy pursued by city leaders. They observed that the city's use of policy tools, such as their police powers and land use authority, were the most likely to cause intense conflict with state regulators. City lawmakers also utilized information disclosure and other persuasive tools to effectuate policy change. These approaches were less likely to elicit as strong of a response from the state. The different 'strategies' are summarized as follows:

- The city's regulatory tools included ordinances and zoning schemes to limit the impacts of air pollution. Each, according to the authors, contributed to the most hostile relationship with industry and state policymakers.
- The city's 'data'-driven policy toolbox included shaming tools such as information disclosure programs and monitoring programs. The collected data were then used as leverage to garner political support among state leaders, industry, and citizens. State and industry officials typically viewed these mechanisms as less combative than the city's regulatory programs.
- The final set of tactics utilized persuasive tools and approaches. Here, city leaders would appeal to the public about the importance of air quality and attempt to change the minds of citizens, industry, and state policymakers.

Setbacks and Quality of Life

Intergovernmental disagreement may result when a state and its cities hold differing values, priorities, or preferences. Differences may emerge regarding oil and gas setbacks and other siting issues. Setbacks function as buffer zones by establishing minimum distances between new oil and gas development and existing occupied structures. Because they can limit where drilling may occur, the politics of setback distances can be intense. Industry and its supporters argue that overly generous setbacks lead to economic inefficiencies, fail to improve public health, and lead to greater disruptions to residents by not allowing operators to cluster development/facilities near one another. Those seeking greater setback distances argue that local governments need the flexibility to protect occupied structures, citizens, green spaces, and bodies of water (COGA 2012a). In some urban and suburban communities, industry has equated large setback distances to de facto bans.

Most states establish specific setback buffers, although they vary greatly in the distances established and in the discretion afforded to communities to exceed state-sanctioned distances. Lawmakers in Oklahoma, Pennsylvania, Ohio, Colorado, and Michigan, for example, have supported a uniform set of setback distances between wells and schools, buildings, occupied structures, and water sources (NPR 2013). By comparison, in Texas, cities may not ban development but can determine commercially reasonable setback distances. The setback distances of selected states are presented in Table 3.4.

Setbacks and other local policies are often designed to minimize the impacts of development on local residents, quality of life, and decrease

Table 3.4 Selected State Setback Distances

State	*Building Setback Distance (in feet)*	*Water Setback Distance*
Texas	200 (locals can exceed but can only establish commercially reasonable distances)	No evidence found
Ohio	100–200	50
Pennsylvania	500	300–1000
California	300—Cities and counties may exceed state minimum	100
Oklahoma	No evidence	No evidence
Louisiana	500—wells drilled below 10,000 750 in Haynesville Shale areas	No evidence
Colorado	500–1000	Determined case by case
Michigan	300	300
New Mexico	No evidence	No evidence
Wyoming	350	350
North Dakota	500	Determined case by case
Illinois	500	500

Source: Richardson et al. (2013)

residential property values (which would harm the local government's property tax base). As discussed in Chapter 1, oil and gas production can impact the character of a community through increased traffic, noise levels, dust contamination, venting, and flaring (Jacquet 2014; Kuster 2012; Ladd 2013; Wiseman 2009). If state institutions are not responsive to their concerns or if state policy preempts local land use/control, citizens may turn to their local government.

Emergency Management

Local governments are often the first responders to fires and other emergency situations. Although fires are a rare site on most well pads, a June 2014 well fire in Monroe County, Ohio, and subsequent emergency response does illustrate the importance of effective emergency management planning and why local governments would push for additional information (Arenschield 2014). The blaze ignited when fluid spilled on hot equipment. The fire quickly spread. Local firefighters reported that they did not know what chemicals were burning and had to wait until StatOil officials could provide a list of chemicals before they could start battling the fire. After 30 minutes, firefighters accessed the on-site lockbox, which contained the company's Material Safety Data Sheets. Local firefighters and StatOil's employees also disagreed on who should fight the blaze, with company personnel asking local responders to leave several times before agreeing to allow firefighters to enter the site. The fire also revealed management challenges for local first responders. Without a communications truck, for example, firefighters were unable to use their radios to communicate at the remote site. In fact, they waited nearly seven hours for a radio truck to arrive. Trucks were also unable to spray a special type of chemical-fire foam far enough to reach some portions of the blaze (Arenschield 2014).

Pennsylvanian first responders are also cognizant of the risks of production-related fires. Bradford County's Deputy Emergency Management Director, Skip Roupp, described a major well fire as a one in 1,000 event, but since "we're getting close to two thousand wells . . . if you follow that logic out, we're due for a major blowout at some point" (Roupp as quoted by Detrow 2011). The state's Fire Commissioner, Ed Mann, added that "If we have a blowout [fire], there's no fire department in Pennsylvania that's going to extinguish that fire" (Mann as quoted by Detrow 2011). He continued that Pennsylvania would likely need to bring in an out-of-state team of specialists should there be a major well fire (Detrow 2011).

Finances

Oil and natural gas production via fracking generates new revenues for state and local governments. Yet, even if local levels of government are amenable to development, they may disagree with the state on finances or the costs of development, or how to measure industry's economic impacts. Recent research by Rabe and Hampton (2015) noted that intergovernmental controversies have erupted with how state and local governments share (or do not) the financial rewards associated with unconventional oil and gas production. The authors identify specific causes of conflict as to how revenues are allocated (i.e., percentage to state versus local), the actual rate (for severance taxes or impact fees), how locals can spend oil and natural gas funds, and how states may use dollars to influence local policymaking. A sampling of state variation is offered in Table 3.5 (Brown and Yucel 2013; Pless 2012a).

Recent empirical work by Raimi and Newell (2014) found that in general most local governments saw their short-term revenues exceed, or at least, equal the increases in short-term costs (see also Bartik et al. 2016). They also noted that downturns in gas and oil prices have meant new

Table 3.5 State Revenue-Sharing Finances

State	*Percentage to Municipalities (as share of total production value)*	*Percentage to Counties (as share of total production value)*	*Percentage to Schools (as share of total production value)*
Texas	.2	.7	3.2 (school districts)
	.3 (other local governments)		1.8 (school trust fund)
Pennsylvania	.4 (grants)	.5	0
	.8		
Montana	.1	2.1	2.4 (school districts)
			1.6 (school trust fund)
Louisiana	.3 (municipalities)	.6 (parishes)	.1 (school trust fund)
Colorado	.3	1.8	2.9 (school districts)
	.3 (other local governments)		1.4 (school trust fund)
	.8 (grants)		
Wyoming	.6	1.5	4.8 (school districts)
	.5 (other local governments)		2.5 (school trust fund)
	.04 (grants)		
North Dakota	.7 (grants)	.8	.3 (school districts)
	.5		2.4 (school trust fund)
Arkansas	.2	.4	.9

Source: Raimi and Newell (2014)

financial challenges for local governments, especially for areas that are economically reliant on oil and gas.

Proximity and Context

Today, oil and gas operations rely on horizontal drilling and hydraulic fracking. This combination has allowed development to take place within just a mile or two of residents and homes, as shown in Table 3.6 (Gold and McGinty 2013). In short, the close proximity to millions of residents means that concerns (both real and perceived) related to emergency management, quality of life/setbacks, and finances (covering the costs of development) are often amplified for local policymakers (see Chapter 1 for a more in-depth discussion of costs and benefits).

The local response with oil and gas interact with other factors listed in this chapter, including, but not limited to: the community's inclination towards environmental protection and sustainability, priorities and goals (economic development), the extent to which industry is seen as a good neighbor, and the percentage of homeowners who view development and interfering with their quality of life. Focusing events (i.e., a spill, accident, fire, or a drinking water-well running dry) can exacerbate many of these real and perceived risk perceptions (Birkland 2011). Each may shape the motivation of community organizers or elected city leaders (Bidwell 2013; Davis and Fisk 2014; Fisk 2016; Hamilton, Colocousis and Duncan 2010; Schlosberg 2007; Sherman 2011; Slovic 1987; Tierney, Lindell, and Perry 2001).

Table 3.6 People Living or Who Have Lived Within One Mile of a Well

State	*Population Since 2000 (in millions)*
Texas	6.09
Ohio	2.63
Pennsylvania	1.78
California	1.62
Oklahoma	1.17
Louisiana	1.06
Colorado	0.34
Michigan	0.25
New Mexico	0.19
Wyoming	0.07
North Dakota	0.04
Total	Approximately 15 million

Source: Gold and McGinty (2013)

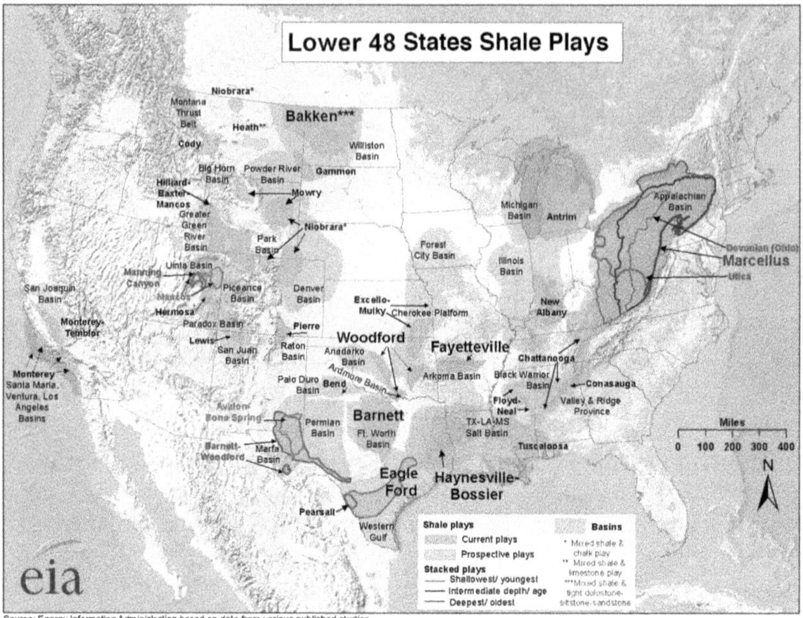

Source: Energy Information Administration based on data from various published studies.
Updated: May 9, 2011

Figure 3.1 Oil and Gas in the United States

The Oil and Gas Landscape of State and Local Governments (as of 2016)

The urban and suburban locations impacted by unconventional oil and gas are spread across the United States (see Figure 3.1).

Most communities' response to oil and gas production is to work within their state's regulatory framework. In short, most communities do not pursue policies or engage in actions that defy the state. However, a small number of cities have defied the state, either symbolically or through the city's substantive/legal authority in each of the states listed in Table 3.7.

An oil- and gas-specific scale is included in Figure 3.2.

Table 3.7 Possible Local Actions (in general)

Scenario	*State-Local Relationship*
Cities possess sufficient expertise and resources to perform state-ordered tasks	Little or no conflict is likely
Cities disagree with state goals, creating an intergovernmental gap, but lack the interest, political support, or willingness to challenge the state	Little or no conflict is likely

Table 3.7 (Continued)

Scenario	*State-Local Relationship*
Cities disagree with state goals, creating an intergovernmental gap, but prefer to use persuasive and other more symbolic measures to challenge the state	Symbolic defiance
Cities disagree with state goals, creating an intergovernmental gap, and are willing to challenge the state's authority typically by refusing or impeding a state policy	Legal defiance

Source: Bowman and Kearney (2011), Fisk (2016)

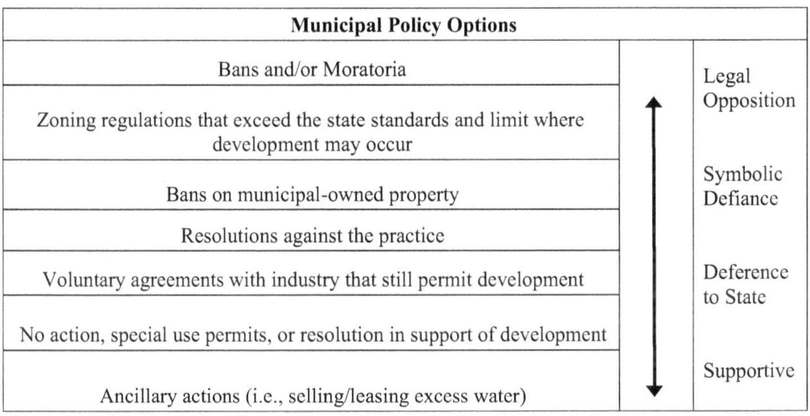

Figure 3.2 Hydraulic Fracturing Policy Action Scale
Source: Fisk (2016)

Colorado and California

Oil and gas exploration have long been part of the fabric of many Western states. However, in recent years a small number of local governments in Colorado and California are pushing back against urban and suburban oil and gas (often relying on fracking and horizontal drilling) operations (Fisk 2016).

Colorado

The state has centralized decision-making authority over well spacing, location, and construction within the Colorado Oil and Gas Conservation Commission (COGCC) and grants local governments a limited role (Colorado Oil and Gas Association 2012a–b; CBS Denver 2012). Nonetheless, the cities of Longmont, Lafayette, Broomfield, and Fort Collins have defied the state through the enactment of bans, moratoria, and other measures designed to restrict oil and gas production. Other

Colorado communities have entered into voluntary agreements with industry designed to assuage the concerns of local leaders and citizens. In response to local defiance, state policymakers have sued the cities of Fort Collins and Longmont, the governor created a task force dedicated to avoiding future state-local disputes, and legislators have considered multiple proposals (Rochat 2012; 2012a).

California

The use of fracking, as part of the extractive process, has also contributed to intergovernmental conflict in California (Veksin 2013). In 2014, several communities banned the practice:

- San Benito County (59 percent supporting the ban)
- Mendocino County (67 percent supporting the ban)
- The city of Los Angeles and Santa Cruz County both have passed bans/moratoria through council action
- In 2016, voters in Alameda, Cruz, and Butte counties banned the technique

(Miller 2015)

In response to its ban, Citadel Exploration filed a lawsuit against San Benito County, claiming that state regulation preempted the county's actions (Miller 2015). In April 2015, however, Citadel announced that it decided to drop suit (Rogers 2015).

Texas, Oklahoma, and New Mexico

Long a part of these three states' economies, oil and gas have enjoyed the support of many state lawmakers. Yet, within Texas, Oklahoma, and New Mexico there is evidence of local government resistance to oil and gas operations (Fisk 2016). Local defiance in Texas and Oklahoma, for example, sparked state legislative action to recentralize regulatory authority in the Texas Railroad Commission and Oklahoma Corporation Commission, respectively. In New Mexico, industry has challenged a local ban in federal court.

Texas

In Texas, cities may regulate many of oil and gas's surface-level impacts, with the Texas Railroad Commission (RRC) maintaining responsibility for its subsurface aspects (Goho 2012). State courts have upheld municipal zoning regulations that limit noise levels for well sites, when deliveries may take place, setbacks, and timelines for repairing broken equipment. In Dallas, for example, the city succeeded in promulgating a setback policy of 1,500 feet (Mosqueda 2013) and in Denton, Texas, residents passed a fracking ban. Other communities, such as Fort Worth and many throughout the Eagle Ford Shale play, are more hospitable to oil and gas development and promulgate lower setback distances (Heinkel-Wolfe and Brown 2012). In response to local activism, state lawmakers passed HB40, which held among other restrictions that local actions may not conflict with state law and that they must be commercially reasonable.

New Mexico

Local pushback is also evident in New Mexico. The city of Las Vegas, New Mexico, expressed support for a statewide moratorium until additional regulations could be enacted. Mora County leaders went further and instituted a countywide ban on oil and gas drilling in 2013 (Cart 2013). Industry quickly filed suit and in early 2015 a federal court ruled that Mora County's ban violated the rights of the Royal Dutch Shell PLC. The court added that the ban destroyed the value of the company's leases and other investments related to tests and collected data (valued at more than one million dollars) (Associated Press 2015).

Oklahoma

No examples exist in Oklahoma of a city or community that has enacted a fracking ban. However, like Texas, Oklahoma state lawmakers passed a 'ban the ban' bill (SB 809) in 2015. The law permits municipalities to enact rules/ordinances that limit oil and natural gas nuisances such as noise, traffic, and odors provided that the rules are "reasonable" (although state and local lawmakers expressed concern about defining what is reasonable) (Henry 2015). Following the bill's passage, the city of Stillwater passed new oil and gas regulations addressing:

- A 660-foot buffer zone for "protected use" properties (homes, schools, parks, churches)
- A reverse setback distance of 400 feet for new structures (homes, schools, etc.) from existing oil and gas wells
- Noise restrictions
- Insurance requirements

A.J. Ferate, the vice president of Regulatory Affairs at the Oklahoma Independent Petroleum Association, described the city's policy as unreasonable and as a de facto ban: "In spirit, it does not allow for oil and gas drilling in city limits" (Ferate as quoted by Wertz 2015).

Ohio, Illinois, West Virginia, and Michigan

Local governments in the region are also defying state lawmakers through bans and restrictive zoning policies. Cities throughout Ohio and Michigan have passed ordinances and zoning policies intended to balance between residents' quality of life and industry's need to efficiently develop oil and gas resources (Negro 2012). However, restrictive municipal policies towards oil and gas production and fracking are not the norm. Voters in Youngstown, Ohio, for example, have rejected several proposed citywide drilling or development bans. Michigan and Ohio communities have also signed lease agreements with developers to drill on public property (Phillips 2012).

Illinois

In 2013, the state passed new rules for firms interested in developing Illinois' shale resources (Southern Illinoisans Fracturing Our Environment 2013; Wernau 2013). Under the new and highly state-centric rules, dissatisfied local governments may request a hearing if they or another organization believes that natural gas development may hurt their community (Redd 2013; Yeagle 2013). Despite little development, state-local tensions are still evident. Mark Haggerty, an analyst with Headwaters Economics in Montana, warned that short-term costs would likely exceed new revenues. He also cautioned that many local governments could because production taxes don't kick in until a well is producing oil, long after a community is beset by transient workers and truck traffic" (Wernau 2013). Five counties in Southern Illinois (Johnson, Jackson, Union, Pope, and Hardin) have already push backed against state policies (SAFE 2013).

Ohio

Ohio lawmakers authorized the state's Division of Mineral Resources Management (part of the Ohio Department of Natural Resources) to oversee oil and natural gas permitting and production throughout the state, even when development takes place inside municipal boundaries. Constitutional provisions, however, charge Ohio's local governments with protecting the public's health/safety/environment and with the authority to pass comprehensive land use plans. The state Supreme Court's jurisprudence in its *Munroe Falls* decision holds that state law preempts local oil and gas ordinances that limit the location of wells or that require a local permit before drilling (Cama 2015; Nolon and Gavin 2013). It is, however, unclear whether the *Munroe Falls* holding applies to municipal zoning policies.

Michigan

Drilling and fracking operations are not as widespread in Michigan compared with Ohio or Pennsylvania. Anxieties over oil and gas development via hydraulic fracking, however, have still contributed to conflicted intergovernmental relations. In an attempt to lead to more efficient development, state lawmakers passed the Michigan Zoning Enabling Act, which centralized regulatory authority as it pertains to the location, drilling, operation, and completion of wells. The legislation also precludes municipalities from adopting zoning schemes that restrict natural gas development (Solomon and Schindler 2012). City and county governments have resisted. The communities of Thornapple, Yankee Springs, Detroit, and Orangeville passed resolutions calling upon lawmakers to "ban fracking to safeguard our citizens from harmful effects and to preserve our environment for generations to come" (Makarewicz 2012, NP). Other Michigan communities have enacted formal challenges to the state. Lawmakers, for example, in West Bloomfield, Rochester Hills, Shelby Township, and Scio Township have promulgated bans and/or moratoria that likely conflict with the Michigan Zoning Enabling Act (Dalbey 2014; Hopkins 2013).

West Virginia

Actions taken by the city council of Morgantown, West Virginia, tested the state-local legal relationship in West Virginia. Following the state's approval of two wells near the city, council members banned drilling within a mile of city limits. Christopher Fletcher, the city's Director of Development Services, explained that the city's goals were not to ban development completely but to "allow them to go where industrial-type development is allowed, and regulate them from a zoning or land-use standpoint as similar as possible to other uses" (Fletcher as quoted by State Journal Staff 2012). The city's legislative actions were struck down by a circuit court.

Louisiana, Florida, and North Carolina

Communities in Louisiana, Florida, and North Carolina are challenging state regulatory approaches to managing unconventional oil and natural gas production. City council members in Bonita Springs, Florida, for example, have enacted a ban on horizontal drilling within city limits. Florida lawmakers have also proposed new laws that would preempt local land use authority to limit oil and gas development.

Louisiana

Conflict is brewing in Louisiana between state leaders and a parish government. The dispute centers on St. Tammany Parish's attempts to restrict oil and gas production that would purportedly threaten its drinking water. The parish sued the State Commissioner of Conservation and the Department of Natural Resources, seeking to block the issuance of a permit.[1] Supporters argue that state control is more efficient and already rests within the statehouse. Opponents counter that local authority in Louisiana includes police powers and zoning authority, and allows the community to protect its drinking water. As of June 2015, Helis Oil and Gas had received a state drilling permit (for a vertical and exploratory well) and a wetlands permit from the Army Corp of Engineers (Warren 2015).

Florida

During the summer of 2015, the city of Bonita Springs joined the city of Coconut Creek in restricting the locations in which hydraulic fracturing may occur. In response, Florida House Republican Ray Rodrigues filed HB 191, which in part reads, "Preempts regulation of all matters relating to exploration, development, production, processing, storage, and transportation of oil and gas." The bill also "declares existing [municipal] ordinances and regulations relating thereto void." Florida Senator Garrett Richter (R) filed a similar bill in the state senate (SB 318). Rodrigues' latest effort is the third attempt to centralize state authority over oil and natural gas production (Sun Bay Staff 2015).

North Carolina

During the 2015 legislative session, North Carolina state lawmakers approved legislation (SB 119) that limited local authority relative to fracking and production. The legislation (SB 119) also invalidates and preempts local actions that exceed state-set standards for operations. The bill passed less than a week after Stokes County approved a three-year moratorium on oil and gas production (Gutiérrez 2015).

New York, Pennsylvania, and Virginia

Despite ongoing economic challenges related to the great recession, intergovernmental disputes have occupied space on policymakers' agendas in each of these states, including lawsuits in Pennsylvania and New York. Throughout each of these states, supportive state lawmakers have described oil and natural gas development as an economic opportunity and lifeline, whereas opponents argue that the development has the potential to ruin the bucolic character of many communities in these states.

New York

Unlike many other states, New York state lawmakers blocked the practice in 2015 (Coin 2015). The ban followed a multi-year environmental and public health study and numerous examples of local bans and moratoria (170 municipalities/counties). Yet, supporters hail the possibility of production as an economic savior for upstate and western portions of the state. Frustrations over the state ban have even contributed to calls for secession by 15 pro-drilling communities. Carolyn Price, Town Supervisor in Windsor, recently described secession as something that has to be examined seriously. She added that, "We only need to drive a few miles [into Pennsylvania], and we can see ourselves . . . gas is the only thing that's truly going to save this area" (Price as quoted by Susman 2015). New York's debate has also included a ruling by the state's highest court that upheld local authority to ban the practice (De Avila, Vilensky, and Gold 2014).

Virginia

Questions relating to local control, intergovernmental boundaries, and hydrocarbon production have also come to the surface in the state of Virginia. State responses have varied. Democratic Attorney General Mark R. Herring, in 2015, concluded that Virginia's local governments have the authority, through their zoning powers, to regulate development, including local bans. Herring's conclusion reverses the opinion issued by his predecessor, Republican Kenneth Cuccinelli. The former attorney general determined that state law permitted communities to enact reasonable restrictions but that they could not ban fracking. The differing advisory opinions are likely to impact thousands of acres, mainly in southwest portions of the state (Dyson 2015).

Pennsylvania

Pennsylvania has been a hotbed of production and of state-local tensions. Under a 2012 law, Act 13, the state required that cities allow drill rigs in all of their 'zones' except for densely populated residential areas. If a city refused, the state would withhold state-collected impact fees. Four communities did not comply, and the state declined to issue their impact fee payments (Phillips 2013, 2012). In response, the cities sued, arguing that the state could not legally withhold the impact fee money. The State Supreme Court found that the portions of Act 13 violated the state constitution (Phillips 2012; National Public Radio 2013; Rabe and Borick 2013).

Focusing on Colorado, Texas, and Ohio

The conflicted intergovernmental relationships described in Chapter 3 represent a wide range of states and regions (Figure 3.3). They include jurisdictions with long histories of oil and gas production to those in which production is fairly new. They also vary along the ideological

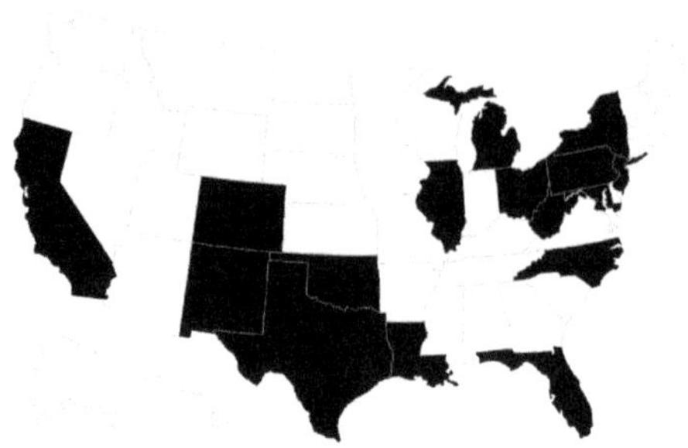

Figure 3.3 Map of Local Defiance

Key: Black—Evidence of local bans, moratoria, or other actions to restrict exploration or production

White—No evidence of local bans, moratoria, or other actions to restrict exploration or production

Source: Author generated

political spectrum from deeply blue to bright red. In addition, they differ to the extent they permit local governments to contribute to oil and gas policymaking, oversight, and enforcement. Each, however, is addressing or has addressed some form of local pushback to unconventional oil and gas development.

Multiple factors may be used to narrow down the number of cases for possible analysis in this project. One possible way is to focus on the distribution of oil and gas's costs and benefits, especially in those areas with a high number of individuals living near development. When production takes place near citizens and within communities, it complicates intergovernmental relationships in a number of ways:

1. States, cities, and citizens are more aware of the benefits ranging from job creation and impacts to their quality of life.
2. States, cities, and citizens may experience the costs unequally such as pollution, spills, and other impacts to residential quality of life.
3. States, cities, and citizens may be home to citizens with different historical experiences, and have differing abilities to access and interpret industry information.
4. City government may become an alternative venue for activists seeking policy change; this can include city council or citizen ballot initiatives.

Colorado, Ohio, and Texas each have a high number of citizens living near oil and natural gas development but differ in their regional location, policies towards oil and gas production, political inclination, and degree to which oil and gas is part of the state's economy (Tables 3.8 and 3.9). This variation offers a number of advantages. First, it helps elucidate the wide variety of challenges to intergovernmental relationships and causes of local defiance. Second, it also expands the 'coverage' of possible solutions based on the ideas that subnational governments are laboratories of democracy. Thus, by examining state-local relations in Colorado, Ohio, and Texas, the best practices proposed in Chapter 9 are novel to some jurisdictions but not others, have already been tested by some subnational policymakers, and are generalizable solutions to the intergovernmental challenges posed by natural resources development.

Table 3.8 Selected State-Level Characteristics for Natural Gas–Producing States

State	*Population Within One Mile of a Site After 1999 (in millions)*	*Natural Gas Production*	*Industry's Percentage of Total GDP (avg. 2008–2011)*
Texas	6.09	8	7.42 percent
Ohio	2.63	20	.12 percent
Pennsylvania	1.78	12	.27 percent
California	1.62	21	.81 percent
Oklahoma	1.17	4	7.86 percent
Louisiana	1.06	2	8.13 percent
Colorado	0.34	7	2.57 percent
Michigan	0.25	19	.05 percent
New Mexico	0.19	3	5.85 percent
Wyoming	0.07	1	15.81 percent
North Dakota	0.04	11	1.96 percent

Sources: (Gold and McGinty 2013; EIA 2014; BEA 2016)

Table 3.9 State Differences and Oil and Gas Development

State	*Proximity to Citizens*	*Natural Break Points*		
		Region	*Politics*	*Oil and Gas Percent of GDP*
Colorado	High	Mountain West	Blue	2.5 percent
Ohio	High	East/Rust Belt	Purple	.12 percent
Texas	High	Oil Patch	Red	7.5 percent

Summary of Possible Sources of Local Defiance

State-local disputes have erupted in a diverse set of states for a variety of reasons. Although local defiance in response to oil and gas development has grown in recent years, it is still a relatively rare event. In New Mexico, for example, of more than 100 cities and counties, only a small handful of communities have engaged in concrete acts of municipal defiance. Reasons for local compliance/cooperation being the norm include the belief that state regulation is sufficient, the realization that substate limitations are inherently risky, and/or the local community is benefitting from production. In Texas, state lawmakers may also enact new legislation that clarifies and/or restricts local authority related to oil and gas. Defiance, on the other hand, is due to a variety of reasons explored in this chapter. These reasons can include differing legal opinions, value differences between state and local officials, proximity to production, anxieties over public safety and environmental harms, finances, and the distribution of benefits. Industry and state leaders, for example, have sued successfully in Colorado and Ohio.

Note

1 The project is proposed by Helis Oil & Gas.

4 Oil and Gas in Colorado

The small number of examples of local defiance in Colorado is a function of two interrelated sets of factors. The first category relates to policy and regulations established by the applicable state agencies. This first category includes questions such as: How does Colorado regulate oil and gas production? What specific regulations address drilling, fracking, and flowback portions of oil and gas development? What authority relative to oil and natural gas production do cities and other local governments possess? How have Colorado lawmakers and courts sorted out related state-local issues in the past? What are the key pieces of legislation, regulation, and court decisions that have shaped state and local relationships? How have Colorado lawmakers utilized less traditional mechanisms (i.e., task forces and direct democracy efforts) to resolve intergovernmental conflicts tied to oil and gas? The answers to these questions represent a critical starting point to understanding local defiance as they establish what state policies local governments are actually pushing back against.

Local defiance is also related to a second category of factors that focuses less on applicable state institutions and more on the place-specific costs and benefits of oil and gas development. This second category answers questions related to: Where is oil and gas development in Colorado? Is development near homes and businesses or drinking water supplies? What communities are benefitting? What communities may be experiencing the negative side effects of oil and gas development? By inventorying costs and benefits of production in terms of location, there may be important clues as to why local defiance is an option pursued by some city governments but not by others.

Following are several reactions from lawmakers to the recommendations put forth by a state-local fracking taskforce. Colorado Governor John Hickenlooper formed the group in 2014 as a compromise that avoided dueling pro- and anti-fracking constitutional ballot initiatives (Jaffe 2015). The selected observations show the range of reactions and the continued frustration felt by some stakeholders over how the state should manage its oil and gas resources:

- La Plata County Commissioner and the Task Force's Co-Chair, Gwen Lachelt, lamented "we let the people of Colorado down . . . by not finding a way to give local governments significantly more control over where drilling happens."

  (Lachelt as quoted by Estabrook 2015)

- The executive director of the Colorado Department of Natural Resources, Mike King, stated that the task force has met its mission to produce several significant regulatory and legislative changes along the lines of what the governor was seeking.

  (King as quoted by Jaffe 2015)

- Governor Hickenlooper described the task force's recommendations as "significant in both breadth and the level of consensus they achieved."

  (Hickenlooper as quoted by Stokols 2015)

- State-local task force member, Matt Sura, was less optimistic about the group's productivity. He suggested that the panel failed in two important aspects. First, the group failed to delineate, clarify, or make any recommendations as to the extent that local governments should regulate oil and natural gas development. Second, the task force did not adequately address the community-level concerns related to the placement and siting of wells, tanks, and other storage infrastructure, and other industry facilities necessary for oil and gas production. He summarized the task force's accomplishments as small enough to "owe Coloradans an apology for not doing better."

  (Sura as quoted by Elliott 2015)

A cacophony of voices have contributed to Colorado's intergovernmental conversations. The "Colorado" conversation elucidates many of the intergovernmental challenges confronting stakeholders in other states and communities. It is replete with lawsuits, conflict, and mistrust, but there is also evidence of effective forms of intergovernmental collaboration, problem solving, and communication. Thus, by understanding what state-level characteristics have led to effective state-local collaboration versus local opposition in Colorado, it is possible to begin larger conversations over what worked and what did not and then apply those best practices in other jurisdictions.

Why Colorado?

Colorado's intergovernmental politics are helpful to other jurisdictions for a variety of reasons. First, through its policies, the state has a history of nurturing oil and gas development. Second, unlike the natural resource booms of the past, advances in horizontal drilling and fracking technology have opened up significant reserves located underneath Colorado's highly populated Front Range communities and in areas of Western Colorado (Sinnott 2012). Third, disputes related to oil and gas production have contributed to a small number of high-profile examples of local defiance, intergovernmental conflict, state-local lawsuits, and several proposed constitutional amendments. Yet, many other local governments work within the state's regulatory framework. Stakeholders have also turned to less conventional policymaking pathways, such as the creation of the state-local task force and a variety of ballot initiatives in hopes of resolving the state's lingering intergovernmental tensions. Given Colorado's diverse set of experiences with the intergovernmental management of oil and gas resources, an important first step to understanding and preventing defiance (Colorado style) is a recognition of the applicable state statues, policies, and state court decisions that address state/local powers, intergovernmental relations, and natural resource management. A second part then turns to Colorado's 'portfolio' of oil- and gas-related risks and rewards.

Colorado's Policymaking Context

Colorado has undergone a political and economic transformation in recent years (Davis 2012). Partially facilitated by the state's former Democratic governor, Bill Ritter, Colorado's new energy economy includes clean and green energy jobs. A political transition is also underway. Since the early 1990s, the state's politics have moderated and have become increasingly liberal, with Democrats winning statewide offices. The environmental movement within the state has also matured and is able to act as a counterweight to industry (Davis 2012; Duffy 2005). Despite the addition of new and greener jobs and revenue sources, Colorado's energy and natural resource community still contributes millions of dollars to state and local economies, employs thousands, and is a key piece in the state's current and future energy mix (Davis 2012). The resulting mixture of politics and priorities has contributed to Colorado being recognized as a leading innovator of environmental policies and also encouraging hydrocarbon production.

Constitutional and Statutory Provisions

Several legal provisions address the scope and character of oil and gas production throughout the state. They also establish the formal parameters and boundaries that govern local control, state authority, and oil and natural gas production.

The state's Oil and Gas Conservation Act of 1951 declared the state's interest in the responsible development of its oil and gas resources (Colo. Rev. Stat. § 34–60–100, et seq.) (Colorado Oil and Gas Conservation

Commission 2012). To ensure this, state lawmakers created the Colorado Oil and Gas Conservation Commission (COGCC) and the oil and gas cleanup fund (Getches-Wilkinson Center for Natural Resources, Energy, and the Environment 2013). The act also established many of the basic rules and priorities that still govern oil and gas development in Colorado. In recent years, lawmakers have revised the state's oil and gas policies in several ways, which are briefly summarized in the following list:

- The state's Air Pollution Prevention and Control Act (Colo. Rev. Stat. § 25–7–100, et seq.) limits the oil and gas industry's impact to the state's ambient air quality and establishes other air pollution standards.
- The Colorado Habitat Stewardship Act of 2007 requires developers to consult with state wildlife officials and surface owners before drilling. It also specifies that operators use best management practices and other reasonable measures to minimize adverse impacts to the state's wildlife.
- Colorado's Water Quality Control Act (Colo. Rev. Stat. § 25–8–100, et seq.) governs the state's water quality and includes provisions related to drinking water quality, construction runoff, and stormwater runoff.

(Getches-Wilkinson Center for Natural Resources, Energy, and the Environment 2013)

- The Clean Air-Clean Jobs Act required that the state's major utilities (Xcel) decommission more than 900 megawatts (MW) of coal-based power generation in favor of natural gas, much of which is recovered through horizontal drilling and hydraulic fracking.
- Colorado voters amended the state's constitution to include renewable portfolio standards and climate change goals.
- The legislature expanded the membership of the COGCC. The revision mandated the inclusion of new commissioners to represent public health, environmental, and wildlife perspectives (Davis 2014).

(More on the COGCC in Section 4.4)

- Colorado is a split-estate state (Colo. Rev. Stat. § 34–60–127), with the mineral estate entitled to reasonable use of the surface to develop the mineral resource. In 2007, the legislature ordered mineral owners to "minimize intrusion upon and damage to the surface of the land," typically accomplished by identifying alternative locations for well sites, pads, or storage ponds/tanks. Operators have also taken steps to reduce and/or cluster production their facilities and gathering pipelines, and have implemented erosion, sediment, and runoff mitigation, and other best management practices.

(Getches-Wilkinson Center for Natural Resources, Energy, and the Environment 2013)

Colorado's basic intergovernmental framework grants home rule authority with several limitations. Cities and towns in Colorado may promulgate and implement zoning ordinances, raise and collect revenues and fees, protect public health and safety, and design comprehensive land use plans and regulations (Col. Const. Art XX). The exercise of municipal home rule authority, however, is contingent on several conditions that may lead to state preemption of local authority:

1. Home rule is only available when the issue or local action includes no extraterritorial impacts (i.e., impacts areas outside city limits) or declared state interests. For issues with mixed state-local impacts, the state's interest will preempt local control or action.
2. City and county governments may not advance policies or ordinances that are arbitrary or capricious.

(*Century Electric Service and Repair, Inc. v. Stone*, 193 Colo. 181, 564 P. 2d 953 1977)

3. Local government regulations may not operationally conflict with state policies or significantly impede a state goal.
4. The application of home rule authority is not guaranteed. State leaders may reduce, clarify, alter, or expand municipal authority as conditions and circumstances dictate, with or without approval from cities and counties.

(*City & County of Denver v. Sweet*, 138 Colo. 41 P. 2d 441 1958)

Extant state law also establishes roles for Colorado's local governments when it comes to oil and gas management. Local governments may appoint a local governmental designee (LGD), who can receive and disseminate information about nearby oil and gas activities. Additionally, the LGD may call for a hearing that evaluates any significant and/or adverse impacts of a proposed well/well pad or a new/modified COGCC rule (COGCC 2008). Cities and counties may also submit testimony during hearings as to localized impacts. They may also regulate roads used by industry to transport equipment and issue general use permits. Finally, state law permits local governments to enter into memorandum of understandings (MOU) with the COGCC and/or industry that articulate alternative standards and commitments. An MOU is a contract designed to clarify the conditions under which an oil/natural gas operator will operate within a particular jurisdiction, and it may be used to supplement or exceed existing COGCC rules (Colorado Municipal League N.d.).

Colorado's local governments have pursued pathways outside of the state's regulatory framework. According to State Official 1 (SO1), "the most significant way that municipal actors have shaped the state's fracking policies is through the local ballot measures that imposed bans or moratoria. Longmont, Fort Collins, Lafayette, and Broomfield have created quite a stir and this has prompted, rightfully so, the current situation with such high stakes [Constitutional ballot initiatives in the Summer of 2014]. Loveland is poised to be next." (Personal Communication, 2014).

State Regulation

Housed within the Department of Natural Resources, the state's major oil and gas regulator is the COGCC, which enacts and enforces most of the state's oil and natural gas regulations, including those related to setbacks, emergency management, and site aesthetics. Lawmakers have charged the COGCC with the responsible development of the state's natural resources in a way that minimizes waste and inefficiencies, protects public health, and safeguards the natural environment (Colorado Department of Natural Resources 2014).

The COGCC has undergone significant changes in recent years. In 2007, the legislature expanded the number of COGCC commissioners from seven to nine (HB-1298 and HB-1341). Before 2007, five of the seven COGCC members represented the industry (typically by professional background). HB 07–1341 broadened COGCC's board membership to include three industry representatives, one member with soil conservation/reclamation expertise, one member to represent landowner/royalty interests, a local government official, an individual who possesses substantial experience in environmental or wildlife protection, and two state officials. The changes, Davis (2012) wrote, contributed to the COGCC passing new rules addressing greater information disclosure, stronger limits on methane emissions, baseline water testing, and increased setback distances from occupied structures.

The COGCC oversees a well's entire lifecycle (Woock 2010). Predrilling activities include registering with the COGCC, submitting requisite forms, impact assessments, permits, and bonds. The COGCC also conducts site inspections and issues enforcement orders (Cook 2015; Davis 2012). A summary list of relevant COGCC regulations is provided in Table 4.1.

The Colorado Department of Public Health and Environment (CDPHE) also impacts the industry via regulations that establish waste, water quality, and air emission standards (Colorado Department of Public Health and Environment [CDPHE] 2014). The CDPHE, through its Water Quality Control Division (WQCD), oversees industry's discharges and issues construction permits. CDPHE's Air Quality Control Commission (AQCC) monitors air quality impacts from oil and gas development. In 2014, the AQCC adopted new rules that required oil and gas operators to conduct regular inspections, capture 95 percent of all fugitive methane

Table 4.1 A Snapshot of Colorado's Fracking Regulations

Regulatory Area	*Colorado*
Pre-drilling water well testing	Required for bodies of water .5 miles from a wellhead.
Water withdrawal restrictions	Addressed in permit.
Casing and cementing depth requirements	50 feet below the water table.
Surface casing cement	Cementing to surface required.
Venting and flaring	COGCC regulations prohibit venting and flaring that is unnecessary or excessive. State regulations also mandate that operators provide notice of flaring to local emergency dispatch or to local governmental designee.
Fluid storage	Wastes may be stored in pits. COGCC regulations require freeboard and liners and five-year tracking.
Underground fluid injection	Allowed.
Disclosure	Current regulation mandates that firms disclose the concentrations of all fracking fluids to state regulators. If fracking fluids are considered a trade secret, the operator must disclose the chemical family to state regulators. In cases of emergency, operators must identify the chemical and its formula to public health officials.
Setbacks	In 2013, the COGCC expanded its setback policies from a 350-foot buffer in urban areas and 150 feet in rural areas to an across the board 500-foot boundary. Operators may drill within 500 feet of an occupied structure in rural areas, if two conditions are met. First, the COGCC director must review and approve the application. Second, the well must incorporate all best-practice mitigation practices (Jaffe 2013a). Other setback regulations restrict drilling/well sites from being located within 1,000 feet of any building that hosts a large number of individuals such as a school or nursing home (Jaffe 2013a).
Seismic activity	Colorado employs a "stoplight" risk management approach, although state regulators have not endorsed any connection between the underground injection of produced waters and heightened levels of induced seismicity (Weaver 2015).

Source: Richardson et al. (2013)

emissions, and make all repairs within 15 days (CDPHE N.d.; Ogburn 2014; Oldham 2014). Finally, the CDPHE regulates industry wastes through its Hazardous Materials and Waste Management Division.

Other Political Efforts

Policymakers seeking to resolve disputes related to oil and gas production have also utilized more unconventional mechanisms. During the first half of 2014, opponents, backed by U.S. Representative Jared Polis, supported Initiatives 88 and 89. Initiative 88 extended Colorado's setback distance to 2,000 feet, whereas Initiative 89 granted citizens a constitutional right to clean air, water, and scenic values (Stokols 2015). The oil and gas industry offered two competing pro-drilling initiatives. Under Initiative 121, local governments, which opted to ban or limit drilling, would lose all of their oil and gas revenues. The second proposal, Initiative 137, required supporters of future ballot initiatives to provide fiscal impact studies and estimates (Tenser 2014).

Seeking to avoid a conflict, Governor Hickenlooper brokered a 2014 compromise, under which both sides agreed to drop their ballot proposals and participate in a state-local task force (Jaffe 2014a). The group's membership consisted of individuals representing industry (oil and gas, agriculture, and home building), local governments and conservation interests, and a variety of professions and backgrounds. The governor ordered the task force to develop recommendations on ways that would minimize future intergovernmental conflicts related to development. The task force was also ordered to identify ways that state and local regulatory systems could be harmonized. In February 2015, the group delivered nine recommendations (supported by two thirds of task-force members) to Governor Hickenlooper (Elliott 2015; Estabrook 2015; Jaffe 2015, 2014a). The recommendations are presented in Table 4.2.

Taskforce member Jeff Robbins reported that the task force considered several recommendations that would have given local governments a larger role in siting issues. Although a majority of taskforce members supported greater forms of local control, many of these proposals did not reach the two-thirds threshold to move forward to the governor (Association Press 2015).

In 2016, activists, many of whom were unsatisfied with the task force's recommendations, again pushed for anti-fracking/local control ballot measures. They settled on Initiative 75, which would have expanded local authority to ban fracking, and Initiative 78, which would have established a 2,500-foot setback (blocking most development inside cities and towns) to be on the 2016 ballot. Both failed to meet the required number of signatures and did not appear on the November 2016 ballot (Williams 2016).

Table 4.2 State-Local Taskforce Recommendations

Proposal	*Implementing Agency*
Create a process to facilitate additional collaboration between operators and local governments, especially if operators are seeking permits for multi-well drill sites.	COGCC
If collaborative processes are unsuccessful at producing an agreement, establish a process for mediation between the parties.	
Require that operators share information and locations of anticipated drilling and production facilities, especially when local governments are engaged in comprehensive planning or if the facility will be large.	COGCC (rules adopted in 2016)
Increase local government designee roles and functions.	COGCC
Expand COGCC staff to include additional inspectors, enforcement, and permitting staff.	Legislature (appropriations), COGCC
Expand staff at the Colorado Department of Public Health and Environment to grow organizational capacities related to health risk assessments and fielding public health complaints.	Legislature (appropriations), CDPHE
Establish an industry information clearinghouse.	COGCC, Legislature (if appropriations are needed)
Establish a working group to study and recommend ways to mitigate industry truck traffic on public streets, roads, and highways.	Multiple agencies
Examine and continue methane rules.	AQCC
Establish a compliance assistance program for oil and gas operators.	COGCC

Source: Jaffe (2015)

Judicial Decisions About Local Control and Oil and Gas

In Colorado, the state courts have often become the final arbiter of disputes involving home rule authority and the relationship between state and local regulation. In effect, their decisions often establish the 'operating' boundaries that are needed for understanding what constitutes an act of local defiance in Colorado. To guide their determinations and decision-making, the court utilizes a multi-part preemption analysis:

1. Is there a need for statewide uniformity?
2. Does the local action affect/impact other jurisdictions (i.e., does the action have an extraterritorial impact)?

3. Is the subject understood as a traditionally state or local power?
4. Does the Colorado Constitution specifically 'name' the state or local communities as the party responsible?

Acts of local defiance related to oil and natural gas in Colorado are not new. In fact, they began in the 1980s and soon migrated their way into the state judiciary. During the fall of 1985, the Greeley City Council and Greeley citizens banned oil and natural gas drilling within the city. The Lundvall Brothers sued the city, arguing that the ordinance conflicted with the COGCC's authority to regulate the location of oil and gas wells, *Voss v. Lundvall Brothers* 830 P. 2d 1061 (1992).

In its decision to nullify the city's ordinance and in favor of the COGCC, the court's reasoning centered on several factors, which have also shaped its subsequent decisions:

1. The state declared an interest in oil and gas development in the Oil and Gas Conservation Act.
2. The state has a compelling interest in enacting regulations that apply uniformly to all operators doing business in the state.
3. The state's oil and gas reserves are located beneath multiple cities and counties, meaning that local regulation would likely have extraterritorial impacts.
4. The COGCC has regulated the oil and gas industry for decades.
5. The ordinance, which blocked oil and gas development, operationally conflicted with state law that encouraged orderly development and thus impeded the state's goals outlined in the Oil and Gas Conservation Act.

As a result, the court voided the city of Greeley's ban, noting "the state's interest in efficient oil and gas development and production throughout the state, as manifested in the Oil and Gas Conservation Act, is sufficiently dominant to override a home-rule city's imposition of a total ban on the drilling of any oil, gas or hydrocarbon wells within the city limits," *Voss v. Lundvall Brothers* 830 P. 2d 1061 (1992); see also Jones 2013. Despite the court's language upholding the COGCC's centralized authority, the decision did reserve some 'room' for local regulations provided that they can be harmonized within the state's regulatory framework.

The State Supreme Court again preempted local regulation in *Bowen/Edwards Associates v. Board of County Commissioners La Plata County, Colorado*, 830 P. 2d 1045 (1992). The dispute centered on a county ordinance that required operators to secure county permits in addition to those issued by the COGCC before drilling. The court described the county's intent as "to facilitate the development of oil and gas resources within the unincorporated area of La Plata County while mitigating potential land-use conflicts between such development and existing, as well as planned, land uses." The court struck down the county permitting program, noting "there is no question that the efficient and equitable development and production of oil and gas resources requires uniform regulation of the technical aspects of drilling, waste prevention, safety precautions, and environmental restoration, [and] also to the location and spacing of wells." *Bowen/Edwards Associates v. Board of County Commissioners, La Plata County, Colorado*, 830 P. 2d 1045 (1992).

Subsequent court decisions have followed the logic outlined in the *Lundvall/Bowen* holdings. Several of these cases are summarized in Table 4.3.

Table 4.3 Local Control Cases

Case	*Analysis*
Town of Frederick v. North American Res. Co. (60 P. 3d 758 [2002])	The court held that state regulation preempted the town's regulatory framework, in several areas: • Setbacks conflicted with COGCC Rule 603a and Colo. Regs. 404–1. • Noise abatement requirements conflicted with COGCC Rule 802. • Visual impacts conflicted with COGCC Rules 318, 803, 804, 1002, and 1003. • The town's penalties against operators conflicted with the COGCC's fine schedule found in Rule 523.
Board of County Commissioners of Gunnison County v. BDS International, LLC, 159 P. 3d 773 (Colo. App. 2006)	The appeals court struck down language that added mitigation fees, required access to company records, and mandated additional bonding based on operational conflicts with state rules (Minor 2014).
Town of Milliken v. Kerr-Mcgee Oil and Gas Onshore LP, 2013WL1908965	The Colorado Court of Appeals ruled that a $400 fee assessed by the town of Milliken on active oil and gas wells located within the community, which funded the town's drive-by inspection program, was unconstitutional. The court held that "oil and gas well site safety and security are matters subject to rule, regulation, order, or permit condition administered by the commission [COGCC]." The court's opinion also noted that "it is irrelevant whether the Commission actually conducts inspections like those performed by the town's police department . . . the relevant inquiry is whether the town's inspection concerned 'matters that are subject to rule, regulation, order, or permit condition administered by the Commission.'"

State courts have also weighed in on more recent examples of local defiance (specifically fracking) following actions taken by the cities of Longmont (a ban) and Fort Collins (five-year moratorium). In July 2014, Boulder County District Court Judge D.D. Mallard nullified the city of Longmont's fracking ban, finding that the city's policy operationally conflicted with the state's Oil and Gas Conservation Act. The opinion noted, "while the court appreciates the Longmont citizens' sincerely held beliefs about risks to their health and safety, the Court does not find this is sufficient to completely devalue the state's interest" (*Colorado Oil and Gas Association, et. al v. City of Longmont, et. al* [13CV63]) (Rochat 2014). With a similar logic, Larimer County District Court Judge Gregory M. Lammons invalidated the city of Fort Collins' moratorium (*Colorado Oil and Gas Association, et. al v. City of Fort Collins, et. al* [13CV31385]) (Handy 2014).

The State Supreme Court, upon appeal, reviewed both cases. In 2016, the justices affirmed the district courts and voided the efforts of the two Front Range cities. Utilizing its preemption framework, the State Supreme Court determined:

1. The use of fracking, as an extraction process, is a mixed issue of state-local concern.
2. The actions taken by cities operationally conflicted with the state's orderly development goals, and thus became an invalid exercise of home rule authority.

(Marmaduke 2016)

3. The opinion also described that even if the court accepted the argument "that fracking is not the only way to produce oil, gas or other hydrocarbons, it appears undisputed that fracking is now standard for virtually all oil and gas wells in Colorado. . . . Longmont's ban, if left in place, could ultimately lead to a patchwork of regulations that would inhibit the efficient development of oil and gas development" (No. 15SC667, *City of Longmont v. Colorado Oil and Gas Association*. 2016 CO 29).

The Oil and Gas Debate in Colorado

Local defiance in Colorado is also shaped by the state's endowment of oil and natural gas resources. The state's reserves, like they are elsewhere in the United States, are not uniformly distributed across the state. In other words, they may be recoverable in one city but not recoverable in an adjacent community. Because of this distribution, the costs and benefits of development are also not evenly felt. Some communities, for example, are more likely to experience significant economic growth, increasing the likelihood that they will be more welcoming of oil and gas wells. Proponents, and in fact, some cities in Colorado, aptly view the development of Colorado's oil and natural gas resources as an economic game-changer and that limits on production would cost the state hundreds of thousands of jobs and billions of dollars in present and future revenues. Other communities and individuals, however, see fewer of the economic benefits associated with production. They may treat the sight of wells as unwanted intrusions that disrupt the character of their communities and threaten their environment.

Oil and Natural Gas Production

Fracking and horizontal drilling have opened up significant shale oil and gas deposits underneath North-Central parts of the state (near Greeley and other Front Range communities) and in Western Colorado near Grand Junction (EIA 2013a) (Table 4.4). Statewide estimates, according to the Energy Information Administration (EIA), totaled 17,139 billion cubic feet (BCF) of dry natural gas and 1,212 million barrels of recoverable oil (EIA 2014, 2014a). In fact, based on proven reserves, the EIA ranks Colorado's Wattenberg field (located along Colorado's Front Range) as one of the largest American oil fields (https://www.eia.gov/state/print.php?sid=CO) (Table 4.5).

Colorado's unconventional-fuels industry benefits from its location and current state policy. Two major gas pipelines cross the state and deliver natural gas to markets in the West and Midwest. The state is also home

Table 4.4 Locations of Reserves and Productive Areas

Field	*Region*
Niobrara Shale (located in multiple states)—	North-Central
Colorado production is centered in the Denver-Julesburg Basin	
San Juan and Raton basins	Southwestern and south-central Colorado (typically produce coal-bed methane)
Piceance Basin	Western Colorado

Source: Davis (2012), EIA (2016b)

Table 4.5 Recoverable Reserves in Colorado (2014)

Resource	*Recoverable Reserves*	*Percent of U.S. Total*
Crude oil	1,212 million barrels	3.8%
Dry natural gas	17,139 billion cubic feet	5.6%

Source: EIA (2016b)

to two major gas-trading hubs and two petroleum refineries (EIA 2014). State policy interventions, such as the Clean Air-Clean Jobs Act, and other factors such as the President's Clean Power Plan have also incentivized gas production in recent years. As illustrated in Figures 4.1 and 4.2, the state's natural gas sector has enjoyed strong and precipitous growth since 2005.

In recent years, operators have also extracted more oil from the state's fields. Production across the state has increased since 2008 (Figure 4.3).

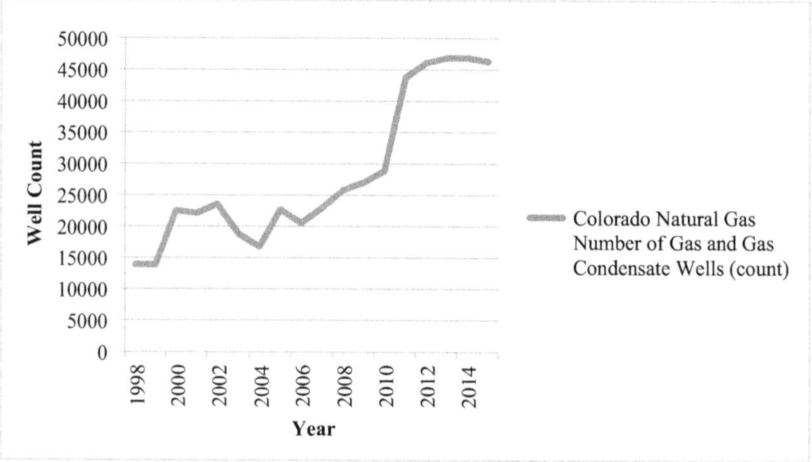

Figure 4.1 Natural Gas Production, by Number of Producing Wells
Source: EIA (2014, 2014b)

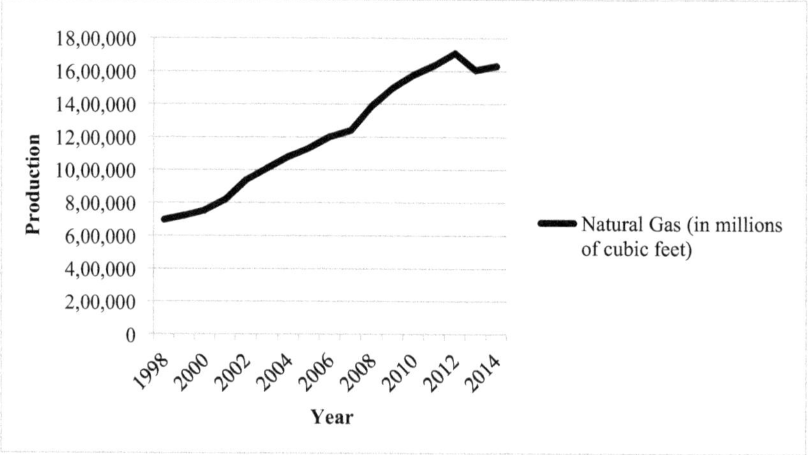

Figure 4.2 Natural Gas Production, by Volume
Source: EIA (2014, 2014b)

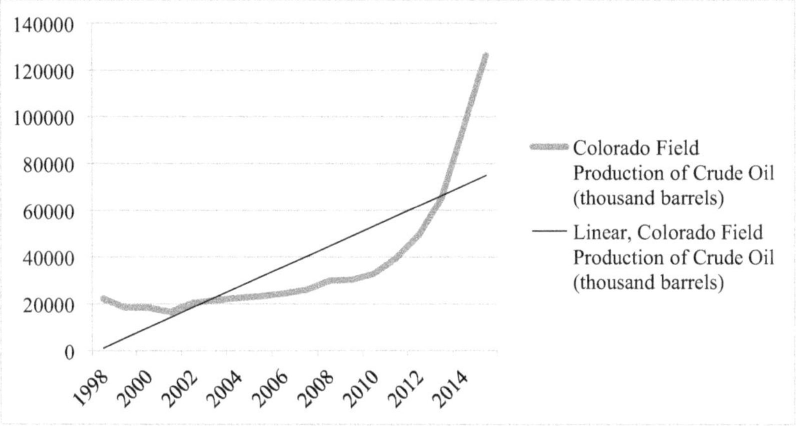

Figure 4.3 Crude Oil Production

Source: EIA (2014c)

The Location of Development

Development in Colorado is increasingly located near significant population centers. Some of this encroachment is due to technological advancements, which have opened up new areas for development. Yet, it is also true that recent growth in Colorado's Front Range has also meant more homes and businesses in areas near oil and gas facilities. Regardless of the 'order,' as of 2013, approximately 340,000 people live or had lived within one mile of an oil or gas well site in Colorado (Gold and McGinty 2013).

The bulk of Colorado's wells are concentrated in and around Front Range counties (Weld) and Western Slope counties (Garfield). Colorado's top-producing counties are shown in Table 4.6 (Jaffe 2014b).

Table 4.6 County-Level Well Numbers (includes both oil and gas)

County	*Producing Wells Since 2011*	*Percent of Total New Wells in the State*
Weld	2,262	49%
Garfield	1,323	28%
Yuma	148	3%
Rio Blanco	109	2%
Mesa	127	3%
La Plata	99	2%
Las Animas	85	2%
All others (includes Front Range counties of Larimer, Boulder, Broomfield, etc.)	506	9%

Source: Randall (2012)

Benefits

The geography behind the state's oil and gas resources impacts the allocation of related costs and benefits. Although specific numbers and assumption are often disputed, oil and natural gas development yields statewide economic dividends. During fiscal year (FY) 2012, for example, Lewandowski and Wobbekind (2013) estimated that the industry generated billions of dollars in direct and indirect economic output and supported thousands of jobs, and was an additional revenue stream for state and local governments. Additionally, oil- and gas-sector employment contributed nearly $6 billion in employee earnings, and millions more in severance taxes; property, sales, and income taxes; fees; royalties; and lease payments. Total estimated economic impacts for FY 2012 are shown in Table 4.7.

Table 4.7 Estimated Oil and Gas Economics in Colorado, FY 2012

	Employee Compensation (millions)	*Value Added (millions)*	*Output (millions)*
Drilling	$319.17	$1,054.36	$1,556.59
Extraction and support activities	$3,942.23	$9,580.08	$18,701.75
Petroleum refineries	$245.89	$1,133.73	$4,789.86
Transportation	$178.85	$263.18	$791.88
Gasoline stations	$466.24	$1,000.47	$1,649.90
All other	$687.31	$1,194.55	$2,078.18
Totals	$5,839.69	$14,226.37	$29,568.16

Source: Lewandowski and Wobbekind (2013)

Employment Benefits

The industry's impact on job creation is a source of conflict. Industry touts that statewide its operations support more than 100,000 jobs, as shown in Table 4.8. Jobs range from those directly involved in oil and gas extraction to secondary and tertiary functions such as support activities, refining, and gas stations (COGA 2012c, 2011). Other estimates place that number around 37,000 (includes those jobs direct-supported by extraction), as shown in Table 4.8 and Figure 4.4.

Since 2013–2014, operators have recalibrated and rescaled their operations. In 2014, WPX Energy, one of the state's largest developers, for

Table 4.8 Oil and Gas Employment Estimates

	Employment
Drilling	4,935
Extraction and support activities	67,572
Petroleum refineries	4,746
Transportation	2,889
Gasoline stations	18,646
All other	12,688
Totals	111,476

Source: Lewandowski and Wobbekind (2013)

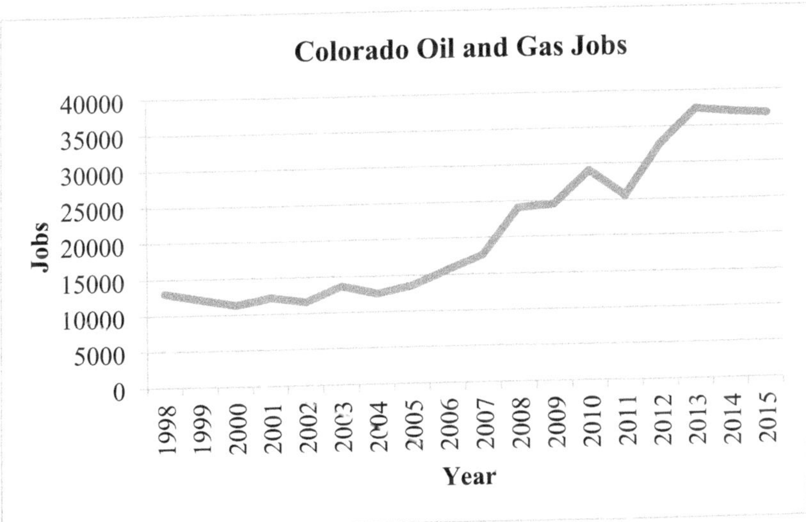

Figure 4.4 Oil and Gas Employment

Note: The BEA defines these jobs as "types of activities included are exploration, drilling, oil and gas well operation and maintenance, the operation of natural gasoline and cycle plants, and the gasification, liquefaction, and pyrolysis of coal at the mine site."

Source: BEA (N.d.)

example, suspended operations at approximately 20 wells in the Piceance Basin (Western Colorado). Encana Corporation, in 2014, also shuttered many of its gas wells in the region (Proctor 2013; Gronewold 2015a). Industry job losses have followed many of the well closures (Lofholm 2015). Colorado Oil and Gas Association (COGA) CEO Dan Haley observed that in 2016 the industry is not experiencing "the layoffs we saw in Colorado in the 1980s, back then, but again we don't know where the bottom is" (Haley as quoted by Wild 2016). As the marketplace for oil and natural gas has weakened, analysts have estimated industry job losses between 2,000 and 6,000 (since 2012 for gas producers and 2014 for oil producers) (Pace 2016).

State Revenue Benefits

Statewide and local revenues from oil and gas production are substantial, even when they are hotly contested (Proctor 2012). Industry's projected fiscal impact for FY 2013 is shown in Table 4.9, with an aggregated economic impact topping $6 billion (COGA 2012, 2012c, 2011; Lewandowski and Wobbekind 2013; Western Natural Resources Law Group 2014).

The state has a financial incentive to encourage oil and gas production. Through the remittance of severance taxes and other revenues, the state receives millions of dollars annually from oil and gas producers (Table 4.10). Much of this revenue originates from a statewide severance tax ranging from 2 to 5 percent, based on the operator's gross annual income (COGA 2012b):

- Gross income <$24,999: tax rate equals 2.0 percent
- Gross income >$25,000 and <$99,999: tax equals $500 plus 3.0 percent over $24,999
- Gross income >$100,000 and <$299,999: tax equals $2,750 plus 4.0 percent over $99,999
- Gross income >$300,000: tax equals $10,750 plus 5.0 percent over $299,999

Table 4.9 Estimated Public Revenues Related to Oil/Natural Gas Production, in Millions (2013)

Severance	*Production*	*Commercial and Residential*	*Leases*	*Royalties*
170.6	357.6	131	61.7	142.5

Personal Income	*Income*	*Sales Tax*	*COGCC Tax*	*Property Tax*
97.8	20.7	138.4	5.9	383.3

Source: Lewandowski and Wobbekind (2013)

Table 4.10 State Severance Tax Collections, by Fiscal Year (FY)

	FY 2004–2005	*FY 2005–2006*	*FY 2006–2007*	*FY 2007–2008*
	$132,315,123	$211,244,521	$116,581,375	$139,910,717
FY 2008–2009	*FY 2009–2010*	*FY 2010–2011*	*FY 2011–2012*	*FY 2012–2013*
$307,003,505	$28,408,847	$130,949,204	$187,132,214	$118,268,697

Source: Colorado Legislative Council Staff (2013)

Once state tax deductions and other exemptions are applied, operators typically pay an effective severance tax rate of between 1 and 2 percent (Colorado Legislative Council Staff 2010, 2008).

- State law permits operators to deduct "87.5 percent of the property taxes paid or assessed on the prior year's oil or natural gas production" from their current year's state severance tax liability. (Colorado Legislative Council Staff 2010)[1]
- Severance taxes are also not collected on wells that produce less than 15 barrels of oil per day or 90 Mcf (thousand cubic feet) of natural gas (also known as 'stripper wells').[2]

Operators also pay a conservation fee to the COGCC (.07% of oil, natural gas, and carbon dioxide production sales, after exemptions have been applied). The fee covers the COGCC's operating expenses and totaled $5.8 million in 2012 and nearly $6 million in 2013 (Lewandowski and Wobbekind 2014).

Local Revenues

Oil and gas revenues also make their way into local budgets, albeit the amount varies and is related to the extent of oil and gas operations in the area (Anderson 2014). Local sources of revenue include taxes received directly from industry (taken as part of operations) and from industry employees (property taxes or sales taxes). They also receive a portion of state severance taxes, with the amount determined by the process depicted in Figure 4.5 (Jaffe 2014b).

The process has yielded economic benefits for impacted counties, with the largest-producing counties often able to cover the additional expenses related to production. For example, during the 'boom's' early years, La Plata County's 2010 budget received approximately $30 million from oil and gas sources (COGA 2012b) (Table 4.11).

Communities experience the costs and benefits of oil and gas unequally, which may help explain varying levels of local support for the industry. In FY 2013, for example, the oil and gas industry paid nearly $265 million in taxes to Weld County, its cities, and school districts (Jaffe 2014b). It also employed thousands of individuals in the county during FY2013. A Weld County official extolled the financial benefits, and noted that industry revenues have enabled the county to achieve a number of its financial goals, including property tax relief and debt retirement: "Weld County is debt free. It's one of the only counties in the country that can say this with the straight face. This is due to solid fiscal planning but is also a testament to the [role that] oil and gas play[s]. Revenues from oil and gas help keep property taxes low. We also don't have a sales tax. Money received from

State Severance Tax	DOLA Portion of Federal Mineral Lease
50% to Department of Natural Resources *50% to Department of Local Affairs*	40% *(of non-bonus payments to state)*
DIVISION OF LOCAL AFFAIRS ↓ ↓	DIVISION OF LOCAL AFFAIRS ↓ ↓
30% Direct Distribution | 70% Grants & Loans	50% Direct Distribution | 50% Grants
↓	↓
County Pool Based on statewide share of the following factors: 1. Colorado Employee Residence 2. Mining and well permits 3. Mineral production	**County Pool** Based on statewide share of the following factors: 1. Colorado Employee Residence Reports 2. Federal Mineral Lease revenue generated
↓	↓
Sub County Pool County/municipality factors: 1. Colorado Employee Residence Reports 2. Population 3. Road miles	**Sub County Pool** County/municipality split factors: 1. Colorado Employee Residence Report 2. Population 3. Road miles

Figure 4.5 Oil and Gas Revenue Allocation Process

Source: Colorado Department of Local Affairs (2015, 2015a)

Table 4.11 Tax Dollars and Local Governments

County	*Oil/Gas Employees*	*Oil/Gas Severance Taxes—From Colorado Department of Local Affairs FY 2014*	*Oil and Gas Production Property Taxes—For All Local Taxing Units (within county) (2012)*
Weld	5,395	$5,045,517.64	$167,004,306
Mesa	2,505	$2,342,728.77	$10,506,632
Adams	1,480	$1,384,127.18	$5,738,353
Larimer	940	$879,107.80	$941,085
Garfield	831	$777,168.70	$85,833,035
Denver	512	$478,833.18	$0
Jefferson	439	$410,562.05	$0
Rio Blanco	398	$372,217.98	$17,123,651
Arapahoe	386	$360,995.33	$666,428
Boulder	366	$342,290.91	$2,617,766
Las Animas	310	$289,918.53	$6,495,872
Morgan	228	$213,230.40	$ 685,512

County	*Oil/Gas Employees*	*Oil/Gas Severance Taxes—From Colorado Department of Local Affairs FY 2014*	*Oil and Gas Production Property Taxes—For All Local Taxing Units (within county) (2012)*
La Plata	218	$203,878.19	$27,619,444
Douglas	167	$156,181.92	$0
Montezuma	146	$136,542.28	$13,981,529
Logan	130	$121,578.74	$1,220,665
El Paso	104	$97,262.99	$0
Yuma	99	$92,586.89	$5,839,822
Moffat	68	$63,595.03	$5,761,811

Source: Colorado Department of Local Affairs (2015, 2015a), Wobbekind, Lewandowski, and Chung (2014)

oil and gas also helps school districts, fire districts, library districts, parks and recreation, roads, streets, bridges, etc."

The county official also pointed to the effects that development has had on property values. The official continued, "Property values continue to go up, not down. Surveys have shown that people that live in Weld like the quality of life. We also have another huge industry in Weld and that is the agricultural industry. Oil and gas has found a way to achieve a harmonious relationship with the Ag industry." Although as of 2016 there was no evidence of local defiance in Weld County, if there were, it could imperil a significant and substantial portion of Weld County's way of life. By comparison, communities without as many economic benefits related to production, such as areas with a greater number of 'clean or green' energy jobs, may view a decision to restrict oil and gas development as less risky and one with fewer direct economic implications.

Costs

Like it does elsewhere in the United States, oil and gas production has a considerable environmental impact. Environmental groups argue that the footprint includes thousands of consumed acres, additional greenhouse gas emissions, water contamination and other particulate emissions. Impacts, however, also vary based on the geological composition of the formation, stage of production (i.e., fracking or flowback operations), and the operator. Table 4.12 presents additional environmental indicators from data compiled by Environment America, an anti-fracking advocacy group, many of which are disputed by industry.

Table 4.12 Environmental Impacts of Fracking

Environmental Harm	*2012 Impact (unless noted)*
Producing wells	32,371 (2013)
Acres damaged since 2005	57,000
Greenhouse gas emissions (based on well completions from 2005 to 2012)	23,000,000 metric tons of carbon dioxide-equivalent
Particulate matter emissions	1,100 tons
Nitrogen oxide	14,000 tons
Carbon monoxide	21,000 tons
Volatile organic compounds	2000 tons
Sulfur dioxide	50 tons

Source: Ridlington and Rumpler (2013)

Empirical research is beginning to unpack the industry's 'footprint' for specific regions and for specific portions of the production process. A 2016 study does suggest the importance of developing models for specific regions/counties, basins, or other sub-units. The study (Salley 2016) noted that:

- Emissions are greatest during flowback processes rather than drilling or fracking.
- Emissions in Western Colorado contained heavier-weight compounds and were slightly higher compared with emissions in the Northern Front Range.

Air Quality Impacts

Opponents argue that emissions released during oil and gas production contribute to climate change and worsen air quality (Jaffe 2014). In May 2012, a University of Colorado and NOAA research team flew over the Denver-Julesburg region and found the following:

- The 20,000 oil and gas wells generated about 13.1 tons of volatile organic emissions per hour.
- Benzene emission rates were captured at levels seven times higher than the CDPHE's estimated rate of 380 pounds/hour.

Citing concerns related to climate change and smog, in 2014 the Air Quality Control Commission (AQCC) announced new rules governing emissions. According to the AQCC's estimates, the rules will reduce smog forming volatile organic chemicals by 92,000 tons and will cut industry's methane emissions by approximately 65,000 tons annually (Jaffe 2014c; Finley 2014c).

Water Supply

Oil and gas production in Colorado complicates the state's water politics (Grantham 2011, Healy 2012). Through a combination of policy instruments and innovations, industry's appetite for water is decreasing. Under a 2012 rule, for example, the COGCC began requiring producers to disclose the total volume of consumed water and the amount of recycled water they plan to use for each well (Rule 205A). New practices related to fluid recycling, identifying operational efficiencies, and recovering other non-potable water have made it possible for operators to recover up to 90 percent of fluids and waters from some wells (this has occurred for a portion of wells drilled in the Piceance Basin). However, because of differing geological formations that necessitate the use of heavier gels and other economic factors, operators extracting in the Denver-Julesburg basin have only recovered 10 percent of used water and other fluids (Jaffe 2014).

Summary on Colorado's Intergovernmental Issues

For Colorado, the boom has brought economic benefits and environmental anxieties, unequally felt across the state. For the state and for regions such as Weld County and in areas of Western Colorado, oil and gas production has contributed significantly to local economies. Despite revenue sharing and some of the most progressive environmental regulations, many skeptics remain unconvinced that oil and gas production via fracking is environmentally benign. These clashes have contributed to the state's fiery fracking-related politics, which have included intergovernmental lawsuits, ballot initiatives that would have effectively ended oil and natural gas production, and other ballot proposals that would have withheld millions of dollars from communities that regulate oil and gas production.

State statutes, court decisions, and regulations have kept the state (and the COGCC) the dominant intergovernmental institution while permitting a small local policymaking role. Yet, the questions of local control are hardly settled and the state is primed for additional attempts to effectuate intergovernmental policy change. These proposals are very likely to involve language that would effectively end most oil and gas development or which severely curtail local land use authority. Given these challenges and options, there is ample room for new and more effective regulatory solutions, which are addressed in Chapter 9.

Notes

1 According to state estimates, the average annual value (for FY 2003–2013) of the ad valorem tax credit was approximately $177 million (Colorado Legislative Council Staff 2013).

2 In 2012, the stripper well exemption applied to 50 percent of the state's natural gas wells and 69 percent of oil wells (Colorado Legislative Council Staff 2013). Colorado Legislative Council staff (2013) estimated the annual value of this exemption for FY 2003–2008 at $89 million.

5 Oil and Gas in Ohio

To better understand local defiance 'Ohio style,' a number of factors are relevant, including statewide characteristics, the geography of development, and the values of local communities. The first portion of this chapter summarizes many of the applicable state laws and administrative rules, how state courts have adjudicated prior disputes and how they further shape oil and gas production in Ohio. It addresses questions such as: What specific regulations address horizontal drilling? What rules apply to portions of the production process that involve fracturing underground formations and flowback? Equally important, what land use authority does state law grant local governments, including cities and townships? Are issues related to fracking, drilling, or the production process a new topic for legislators, regulators, and the state courts? If not, how does the state's history and previous rules inform current oil- and gas-related disputes? What are the critical pieces of legislation, regulation, and court decisions that have shaped (and are still shaping) intergovernmental relationships? Have activists looked to new venues such as ballot initiatives or other unconventional venues to bring about policy change? The answers to these questions form the 'state' side of local defiance .

A common theme is that the politics of oil and natural gas production are often place-specific. In Ohio, this is certainly true. As such, part of understanding the sources and motivations behind local defiance also includes a familiarity with oil and gas's distribution of costs and benefits within Ohio. Is development impacting the state and its local governments equally? If not, who is benefiting and to what extent are costs concentrated or spread among entire regions? What are some of the concerns reported by local governments and other groups skeptical of production? These answers also inform intergovernmental conversations centering on how to improve state-local governance of oil and natural gas production.

Ohio's intergovernmental battles offer insights into what is meant by local democracy and governance, which are useful in the search for more effective regulations and policy approaches. Following are a sample of reactions to the Ohio Supreme Court's decision that preempted local oil and gas ordinances (it is unclear if the court's ruling also included local zoning policies):

- Shawn Bennett, the Executive Vice President of the Ohio Oil and Gas Association, described that oil and gas development "is a matter of statewide interest and should be managed by professionals with the expertise to adequately regulate and oversee the industry."
(Bennett as quoted by Balmert 2015)

- The mayor of Athens, Ohio, Paul Wiehl commented "I guess that means the voice of the people doesn't matter . . . we said—we want local control,' and then they take it away."
(Wiehl as quoted by Bohatch 2015)

- Ohio Secretary of State John Husted suggested, "If advocates of these local proposals truly want to effect change, they should use the legislative process available to them" (meaning elected state officials).
(Husted as quoted by Carr-Smyth 2015; Dennis 2015)

- Governor John Kasich described the importance of the state's responsibility in blunt economic terms, "If we deny ourselves the ability to develop reserves, we will be shutting down the hopes and dreams of a lot of children . . . and families"
(Kasich as quoted by Wang 2011).

Why Ohio?

Ohio's intergovernmental experiences related to oil and gas production are instructive for other jurisdictions seeking to identify more optimal governing arrangements. First, the Marcellus and Utica Shale plays are underneath large portions of Eastern Ohio. Their development has excited state and local policymakers and has contributed to a renewed sense of optimism for many of the region's battered local economies. Development in many of these locations has also prompted concerns for many environmentalists, related to water contamination and earthquakes. Second, the 'boom' has brought production closer to many communities, including Youngstown, Canton, and suburban areas located near Cleveland and Akron. By 2013, more than two million Ohio residents lived or had lived within a mile of a well (this includes all oil and gas wells—not just those drilled since 2006) (Gold and McGinty 2013). Finally, Ohio's fracking travails and 'story' include a state-local lawsuit, several failed attempts by activists to ban fracking locally, and a widely publicized dispute over severance tax rates and local revenue sharing between the Republican governor and Republican lawmakers.

Much like Colorado and Texas, the 'Ohio' story adds another voice to the intergovernmental conversations that center on oil- and gas-related governance. Stakeholders here have engaged in defiance and have conflict but also have utilized inventive and innovative collaborations to try and balance state and local concerns. As a result, there are lessons to be learned by understanding Ohio's shale story.

Ohio's Policymaking Context

Both political parties have supported unconventional oil and natural gas development. In 2010, for example, a bipartisan group of lawmakers passed SB 165, which included new environmental standards, protections for urban communities, and new requirements for developers. Political support for oil and gas drilling continued under Republican Governor John Kasich. Under the Kasich administration, Ohio lawmakers passed SB 315, which established new standards for production, processing, and waste management. Although state leaders have differed on the scope of the costs and benefits attached to shale production, they have noted the statewide economic potential of developing Ohio's oil and gas industry.

Despite bipartisan support for shale development, state Republicans have disagreed over increasing the state's severance tax and the extent to which the state should share its oil and gas revenues with local governments (Kovac 2013). The latter issue is important as local governments often seek out state aid to help them offset many of the costs associated with oil and gas development (operators already pay property taxes to local governments on the estimated value of their reserves). Three of the proposals are included in Table 5.1, with a focus on tax rates and the degree of state-local revenue sharing.

Unable to resolve the severance tax question, lawmakers turned to a legislative study group. The group agreed with Governor Kasich that severance industry's tax rate was comparatively low at 10 cents per barrel of

Table 5.1 Competing Intergovernmental Proposals

	Governor Kasich's 2014 Plan	*Governor Kasich's 2015 Plan*	*House Republicans—HB 375*
Severance tax rate—(oil and gas sold at wellhead)	2.75% severance tax	6.5% severance tax and 4.5 percent charge to natural gas liquids	Approximately 1% (based on the gross receipts collected during the well's first two years of production); after two years, the tax rate would increase to 2.25%
State revenue sharing	25% of state severance tax collections	20% of state severance tax collections	17.5% of the collections (distributed through an 11-member Shale Gas Regional Commission)

Sources: Dolan (2015), Funk (2014), McParland (2014), Kovac (2013), Pelzer (2014), Vardon (2013)

oil and 2.5 cents per 1,000 cubic feet of natural gas. However, the group also acknowledged that depressed oil and gas prices have contributed to financial hardships across Ohio's oil and gas community, making it more difficult to support the governor's tax plans that would increase the severance tax rate. The study group also recommended the use of tax triggers, incremental changes in tax rates, and increasing assistance for affected counties (Pelzer 2015).

Constitutional and Statutory Provisions

Applicable language in the State Constitution and statutes provide useful guideposts to understanding Ohio's intergovernmental challenges. Together, they establish the legal boundaries for what constitutes local defiance, what is required of the oil and gas community, and what are permissible actions for local governments.

Article XVIII, Section 3, of the Ohio State Constitution defines the extent of local lawmaking authority. The Article empowers the state's local governments "to exercise all powers of local self-government and to adopt and enforce within their limits such local police, sanitary, and other similar regulations" *Morris v. Roseman*, 162 Ohio St. 447 (1954). The State Supreme Court has placed a number of relevant caveats that limit the exercise of home rule authority:

1. When local actions have extraterritorial impacts
2. When local laws operationally or materially conflict with applicable state general laws
3. When local laws are clearly state roles as defined by the court or state statute
    a. Previous court decisions have clarified state and local roles, as shown in Tables 5.2 and 5.3.

(Gotherman 2002)

Table 5.2 Court-Declared Local Powers

Case	*General Power*
Angell v. Toledo (1950) 153 Ohio St. 179	The power to tax
Gordon v. Rhodes (1952) 158 Ohio St. 129	The power to incur debt
State Ex rel. Brustle v. Rich (1953) 159 Ohio St. 13	The power to make urban renewal decisions (includes eminent domain)
Fitzgerald v. Cleveland (1913), 88 Ohio St. 338; *Mansfield v. Endly* (1931), 38 Ohio App. 528	The power to make personnel and governmental structure decisions

Source: Gotherman (2002)

Table 5.3 Court-Declared State Powers

Case	*General Power*
Bucyrus v. Dept. of Health (1929), 120 Ohio St. 426	The power to establish statewide sewage treatment standards
Cleveland Electric Illumination Co. v. Painesville (1968), 15 Ohio St. 2d 125	The power to establish cross-county electric transmission lines
Evans v. Moore (1982), 69 Ohio St. 2d 88	The power to set prevailing wage laws
Kettering v. State Emp. Relations Bd. (1986), 26 Ohio St. 3d 50	The power to establish parameters for labor relations

Source: Gotherman (2002)

The state's property laws also influence Ohio's intergovernmental oil and gas politics. Article 1, Section 19b, of the State Constitution describes water as a property right. State rules apply if a property owner's use of water interferes with other users, as defined by Section 1521.16 of the Ohio Revised Code. If triggered, state rules mandate the issuance of a permit from the Ohio Department of Natural Resources' (ODNR) Division of Soil and Water Resources (Division of Soil and Water Resources [DWSR] 2012).

Like Colorado and Texas, a number of other state laws impact environment quality and energy production in Ohio, as identified in Table 5.4.

Table 5.4 Major Environmental and Energy Laws

Name and Year	*Policy Area*	*Agencies*
Ohio Solid and Hazardous Waste and Disposal Act (1967)	Reducing and/or preventing pollution	Division of Materials and Waste Management
Clean Air Act (1970)	Stationary and mobile air sources and state implementation plans	Ohio Environmental Protection Agency (OEPA)
Clean Water Act (1970)	Regulate emissions from wastewater treatment plants, factories, and stormwater to reduce the impact of pollutants on the state's water resources	OEPA
Renewable portfolio standard	Requires 12.5% of the state's power come from renewable sources by 2027	Regulated utilities
Energy efficiency standard	Requires 22% reduction in retail electricity sales by 2027	Regulated entities

Source: Ohio EPA (2014a)

The Ohio Oil and Gas Act of 1965 is the state's main oil and gas framework. Between 2010 and 2015, state lawmakers revised the act to include a number of additional provisions to encourage production while also improving health and safety:

- Added statewide standards on well construction, permits, operations, and wastewater storage
- Added new provisions to acquire a drilling permit, including notification, time, spacing requirements, fees, and inspection schedules
- Added language that requires operators to submit water consumption data, including estimated rates and volumes
- Increased the insurance rates and fees charged to operators
- Established mandatory daily fines
- Authorized the hiring of additional enforcement agents

Legislative updates have also shaped subsequent intergovernmental interactions (McParland 2014, 2014a):

- Designated the ODNR as "the sole and exclusive" oil and gas regulatory authority
- Established statewide minimum setback distances for wells, tank batteries, and other oil and gas surface facilities from an occupied structure and other buildings
- Restricted new directionally drilled wells from being closer than 150 feet to any property line unless the operator has secured the landowner's written consent
- Clarified that local government may call a hearing relative to oil and/or natural gas wells or facilities
- Required that new surface wells cannot be located closer than 100 feet to an occupied dwelling or public building
- Established new permit fees for urban areas

(ODNR 2016; Simmers 2012)

State Regulations

The Ohio Environmental Protection Agency (OEPA) and the Ohio Department of Natural Resources Division of Mineral Resources Management (ODNR-DMRM) co-manage oil and gas operations. The OEPA monitors air and water emissions, whereas the ODNR regulates drilling operations, production, and waste disposal (ODNR 2014; Shale 2014; Sandrick 2015a).

State lawmakers created the OEPA in 1972. The agency is organized by various divisions (air, water, hazardous wastes, and site remediation divisions) that issue applicable permits or licenses, conduct enforcement actions, and inspect facilities. Divisions may also provide technical assistance to regulated entities (OEPA 2014). Currently, the OEPA requires that oil and gas operators receive a permit-to-install and operate (PTIO) before engaging in activities that emit air pollutants (unless exempted). According to OEPA, an Ohio drill site may include several potential sources of air emissions: dehydration systems, engines, unpaved roads, storage tanks, generators, pipelines, and combustion devices. The OEPA also regulates the disposal of contaminated soils, drill cuttings, muds, and the sands used during production (OEPA 2014a).

The Ohio Department of Natural Resources (ODNR) oversees many of the subsurface and technical aspects of oil and gas production. The ODNR's permitting process follows a fairly standard protocol, as discussed in Chapter 1 (ODNR 2014). Other pertinent regulations for drilling and fracking are included in Table 5.5.

Table 5.5 Ohio's Regulations

Area	*Regulation*
Pre-drilling water well testing	Within 1500 feet of well
Water withdrawal restrictions	Permit required if >2m gal/day
Venting and flaring	Banned, restricted
Fluid storage	Permit required for all pits and tanks
Underground fluid injection	Permit/approval and record-keeping required
Disclosure	Current regulation mandates that firms disclose fracking fluids to FracFocus—this includes (if applicable) the type, volume, and concentration of acid, and the type and volume of fluid used to stimulate the well's reservoir. In the event of a spill, the ODNR may obtain proprietary chemical formulas from operators.
	Physicians may share trade-secreted chemical information with patients and other medical personnel if they are directly involved in patient treatment (ODNR 2015).
Setbacks	Recent amendments to the Ohio Oil and Gas Act Chapter 1509 expanded setbacks to 150 feet—unless approved by the property owner and Chief of the Division of Mineral Resources Management.
Seismic activity	Ohio employs a "stoplight" approach, which has resulted in moratoriums being placed on several seismically active wastewater wells (Brown 2015).

Source: Richardson et al. (2013)

Table 5.6 Co-Regulators

	Ohio Department of Natural Resources	*Ohio Environmental Protection Agency*
Drilling	• Issues drilling permits • Establishes requirements for spacing/location, design and construction of wells • Performs inspections and oversight • Sets requirements for spills, releases, and well plugging/abandonment • Issues permits to withdraw water	• Mandates that operators receive authorization for construction when a proposed activity impacts wetlands, streams, rivers • Drillers must obtain an air permit-to-install and operate
Wastewater	• Creates and enacts design requirements and closure requirements when operators plan to store drill cuttings and brine/flowback water in on-site pits or lagoons • Establishes standards for cuttings and sediments left on-site	• Once soil is considered a solid waste, operators are required to follow solid waste protocols for shipping it off-site
Water disposal	• Regulates the disposal of brine and sets design standards and oversees operation of underground injection wells • Responds to citizens	

Source: Shale Info (2014, Ohio 2014b); OEPA (2014b)

As co-regulators, the OEPA and the ODNR policies shape the environmental footprint of the oil and gas industry. They also afford stakeholders additional venues to push for their preferred policies related to oil and gas development. Table 5.6 summarizes these rules based on the activity.

Local Involvement

Lawmakers, affirmed by the State Supreme Court, have preempted most local oil and gas ordinances. Local governments (townships) may enter into road-use maintenance agreements (RUMAs) with oil and gas operators, participate in ODNR rulemaking hearings, and issue construction and right-of-way permits (as long as they are not issued in an arbitrary or discriminatory way). RUMAs typically address the following:

- Establish responsibility for road construction, maintenance, and repairs—locals may seek to be completely compensated for costs to repair and maintain roads impacted by oil and gas operation
- Establish designated travel routes for industry trucks, tankers, etc., to minimize road impacts
- Local authorities may regulate the actual spreading of fluids on applicable roads. These may be more stringent that state standards.

(ODNR 2015)

Under the State Constitution, local governments also have the authority to conserve and protect their natural and open spaces. Nolon and Gavin (2013) write that a handful of communities have used this authority to create Environmental Quality Districts (in Cincinnati) and an Open Space District in Rootstown Township. State courts have upheld constitutionality of these districts, explaining that they are a valid exercise of local authority to protect public safety, health, and welfare.

Judicial Decisions About Local Control and Oil and Gas

Ohio's Supreme Court utilizes a home rule/preemption analysis as a template to resolve intergovernmental conflicts:

1. A court determines whether the action in question is an exercise of local self-government or an exercise of local police power.
   a. If an exercise of local self-government, the home rule analysis stops because a municipality is granted such power under the State Constitution.
2. If the action involves the use of a concurrent police power (legislation that protects public health, safety, morals, or residents' general welfare), then the court applies a secondary analysis to ascertain whether the state statute is a general law. The court's general law analysis involves four 'questions':
   a. Is the statute part of a statewide enactment and is it comprehensive?
   b. Is the statute applied uniformly throughout the state?
   c. Does it establish police, sanitary, or other standards that are not just restrictions on state power?
   d. Does it articulate rules of conduct upon citizens?
3. Finally, the court then applies its conflict test. Here, members of the judiciary evaluate whether the state general law invalidates local policies by assessing if there are any inconsistencies, contradictions, or operational conflicts between state policy and the local ordinance.

(*Rispo Realty & Development Co. v. Parma*, 55 Ohio St. 3d 101 1990)

Table 5.7 State Courts, Intergovernmental Relations, and Oil and Gas

Case	*Dispute*	*Home Rule Analysis and Ruling*
Natale v. Everflow Eastern, Inc., Case No. 2010-T-0088 (OH Ct. App., Dist. 11, Aug. 29, 2011)	Construction of a well on neighboring property constituted a nuisance and was in violation of local zoning ordinances	The court determined that the ODNR *is the sole and exclusive authority to regulate the permitting, location, and spacing of oil and gas wells and production operations within the state.* This interest, according to the court, is of statewide interest that necessitates uniform state regulations. The court held "since appellant's claims are clearly based on the *location* and *operation* of the oil and gas well, it is manifest that the ordinances referenced by appellant are preempted by the clear, unequivocal language in the statute."
Newbury Township Board of Trustees v. Lomak Petroleum, Inc., 583 N.E.2d 302 (Ohio 1992)	Drilling bans in specific areas within a community	Court preempted the township's ban on drilling in residential and farmland zones because the town's leaders failed to demonstrate a compelling public safety/health reason for the ban (Nolon and Gavin 2013).

Recently, the state judiciary was again relied upon to resolve intergovernmental conflicts related to oil and gas development in *State ex rel. Morrison v. Beck Energy Corp.*, 143 Ohio St.3d 271. At dispute was the constitutionality of the city of Munroe Falls' ordinance that regulated oil and gas drilling on private property within the city. The trial court held in favor of the city. The appellate court reversed and concluded that state lawmakers intended the ODNR to be the 'sole and exclusive' regulatory authority within the state and that the city's ordinance conflicted with Chapter 1509 of the state's Oil and Gas Act (Cocklin 2013).

City officials appealed to the State Supreme Court. The justices applied the court's Home Rule Analysis and concluded that the city's actions were taken as part of its police powers. It also concluded that Chapter 1509.02 of the Oil and Gas law was a general law, triggering the court's conflict analysis. In a 4–3 decision, the court nullified the city's ordinance based on two lines of legal reasoning:

1. Beck Energy received a state permit to vertically drill within the city's limits. The court determined that because the city's oil and gas ordinance (which could take more than a year) also required a permit before drilling, the ordinance prevented Beck from engaging in an activity that was expressly permitted by the state.
2. Section 1509.02 of the Oil and Gas Act made the ODNR the "sole and exclusive authority to regulate the permitting, location, and spacing of oil and gas wells and production operations." The legislature, according to the court, authorized local governments to act in policy areas addressing infrastructure, alleys, aqueducts, and road maintenance but not in siting decisions.

(*State ex rel. Morrison v. Beck Energy Corp.*, 143 Ohio St.3d 271 2015)

Following the *Beck* decision, lower courts have nullified additional local oil and gas ordinances. In Broadview Heights, Ohio, for example, Cuyahoga County Trial Court Judge Michael K. Astrab invalidated a local fracking ban approved by local voters in 2012. In his decision, Judge Astrab echoed the Supreme Court's logic that the ODNR is the "sole and exclusive authority" related to well permitting, location, and spacing. Additionally, he found that the ban "discriminates against, unfairly impedes, or obstructs oil and gas activities and operations" (Sandrick 2015).

The Oil and Gas Debate in Ohio

Ohio's intergovernmental and policy debates mirror many other statehouses across the country. Supporters argue that production has contributed and will continue to contribute to a strong and robust state economy, with risks mitigated through the implementation of best management practices. And, much like supportive governors in Colorado and Texas, Ohio's supporters suggest that uniform statewide policies ensure efficiency and avoid the entanglements inherent to a patchwork of local regulations. Opponents shift the debate towards oil and gas's impacts to the environment, public health, and to residents' quality of life. These arguments, similar to arguments made by activists in Colorado and Texas, are often influenced by place-specific factors such as the location of the development, the accrual of economic benefits, and the real/perceived threats to public health, well-being, and the environment.

Oil and Natural Gas Production

Ohio's commercial oil and natural gas industry began in the late 1880s, peaked in 1896, and waned through the first half of the 20th century (Ohio Oil and Gas Association 2013, 2012). By 1951, developers were again interested in Ohio's oil and gas resources although their interest quickly dissipated. The fracking-fueled oil and gas renaissance has sparked another wave of interest in Ohio's Utica Shale. The Energy Information Administration (EIA) estimates that approximately 6,723 billion cubic feet of natural gas and 78 million barrels of oil are located beneath Ohio's farms, fields, townships, and cities (Tables 5.8 and 5.9).

Ohio's location as a distribution crossroads for more expensive gas markets, especially along the East Coast, has also aided production and support for development. To supply markets throughout the United States and Canada, Ohio has experienced a number of new pipeline construction projects. By 2024, analysts estimate that industry will add nearly 38,000 miles of new and improved lines (Chavez 2014). Table 5.10 summarizes five major pipeline projects (as of 2015).

Development in the Utica region has also enabled firms to transport Ohio gas west towards Chicago. In these areas, gas prices have stayed considerably higher compared with areas closer to the Utica and Marcellus shale regions (Funk 2016).

President Obama's Clean Power Plan (CPP) has further shaped demand for Utica shale resources. Two of the state's dominant utilities (American Electric Power and FirstEnergy Corp.), for example, are shuttering coal plants and investing in natural gas and renewable energy. Asim Haque, the chairperson of the Public Utilities Commission of Ohio, summarized the state's two-pronged response to the plan as one in which the state is suing the EPA and "trying to determine what intelligible compliance will look like for the state" (Haque as quoted by Kuckro 2016). Although

Table 5.8 Recoverable Reserves in Ohio (2014)

Reserves	*Ohio*	*Percent of U.S. Total*
Crude oil	62 million barrels	0.2%
Dry natural gas	12,104 billion cubic feet	3.9%

Source: EIA (2016c)

Table 5.9 Location of Reserves

Shale Play	*Region*
Marcellus	Extreme Eastern Ohio (mostly in Pennsylvania, West Virginia, and New York)
Utica	Eastern half of the state

Source: EIA (2013a)

Table 5.10 Major Ohio Pipelines

Company	*Cost (in 2016 dollars)*	*Length (miles)*	*Status*	*Expected Capacity*	*Region*
Energy Transfer Partner L.P.'s Rover Pipeline	4.3 billion	823	Operational in 2016	3.25 billion cubic feet of gas daily	Southeast Ohio to Michigan and Canada
Spectra Energy/ DTE Energy	1.5 billion	250	Scheduled completion 2017	2 billion cubic feet of gas daily	Northeast Ohio to Michigan and Canada
Columbia Pipeline Group	1.75 billion	160	Scheduled completion 2017	1.5 billion cubic feet of gas daily	West Virginia and Southeast Ohio to Central Ohio, where it connects with lines heading to Kentucky
Enterprise Products Partners		1230	Operational 2013	90,000 barrels a day of ethane (a natural gas liquid)	Southwest Pennsylvania through to Southern Indiana
Sunoco Logistics and Mark West Liberty Midstream (operational)		230	Operational	50,000 barrels of ethane per day	Youngstown area to Ontario

Source: Chavez (2014)

Haque did not identify what the state's compliance plan would ultimately be, in a presentation to the Reliability Technical Conference organized by the Federal Energy Regulatory Commission, he did identify that natural gas and renewable power resources are often used to replace retiring coal units (Haque 2015).

The cumulative effect of favorable geography, federal policy, and state regulatory approaches has led to precipitous growth in recent years, as shown in Figures 5.1 through 5.3. Gas production has continued to increase throughout 2015, despite drops in the number of new wells drilled, gas prices, and drilling rigs (Funk 2016). Oil production within Ohio has continued to increase as well.

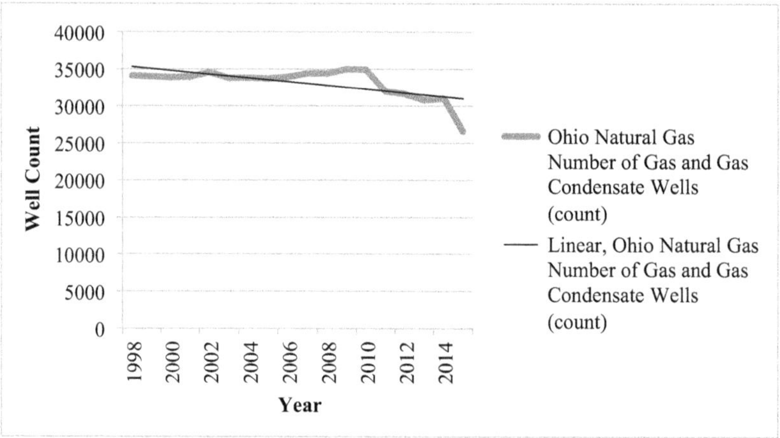

Figure 5.1 Natural Gas Production, by Number of Producing Wells (count)
Source: EIA (2014, 2014a)

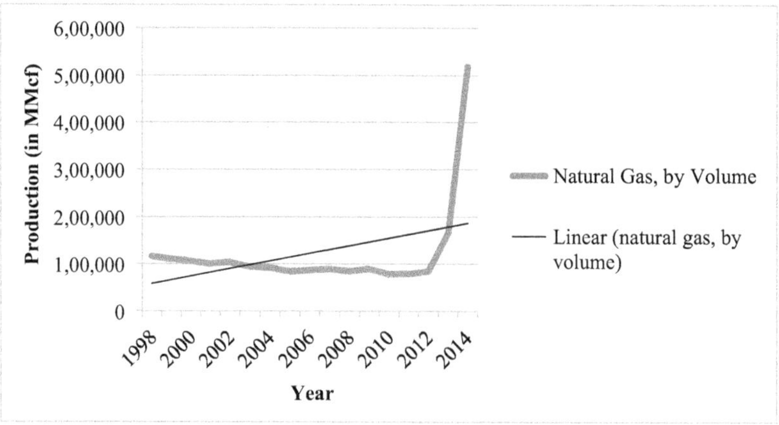

Figure 5.2 Natural Gas Production, by Volume (in millions of cubic feet [MMcf])
Source: EIA (2014, 2014b)

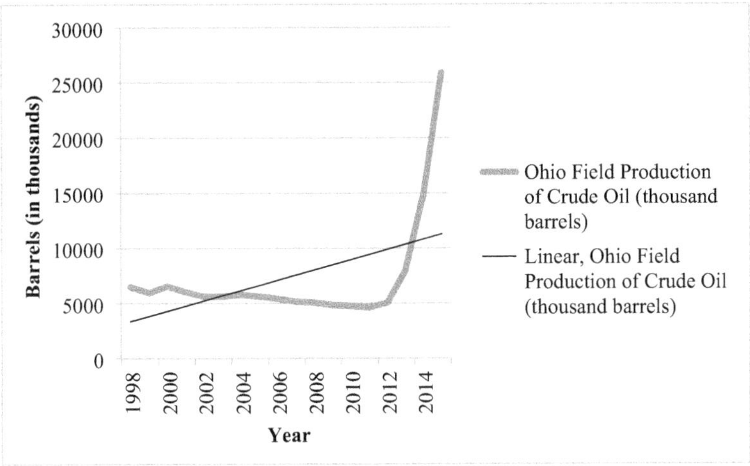

Figure 5.3 Oil Production (in thousands of barrels)
Source: EIA (2014c)

The Location of Development

Like it has elsewhere, the combination of horizontal drilling and fracking technologies have opened significant deposits in Eastern Ohio, including areas near Cleveland, Youngstown, and Akron. By 2013, more than 2.6 million Ohioans lived or had lived within a mile of an oil or gas well (although not all of these individuals live or lived near a well that had been horizontally drilled or hydraulically fractured) (Gold and McGinty 2013). As of 2016, production was centered mainly in several Eastern Ohio locations, with Carroll, Noble, Licking, Knox, Columbiana, Stark, and Monroe counties being the leading areas of production (Tables 5.11 and 5.12).

Table 5.11 Local Production (2012 most active counties, by wells)

County	*2012 Wells*
Carroll	87
Noble	49
Licking	42
Knox	40
Stark	36
Monroe	31
Columbiana	30
Harrison	22
Coshocton	18
Guernsey	18

Source: Simmers 2012

Table 5.12 Local Production (2015 most active counties, by wells)

County	*Percentage of Total Wells*	*2015 Wells*	*Change from 2014*
Harrison	19%	90	53%
Carroll	17%	81	-51%
Belmont	15%	72	148%
Monroe	11%	53	89%
Guernsey	10%	49	53%
Noble	9%	42	24%
Coshocton	4%	17	31%
Stark	3%	16	-47%
Knox	3%	15	-12%
Columbiana	3%	14	-50%

Source: Ohio Oil and Gas Association (2015)

Revenue and Tax Benefits

The oil and gas industry produces a variety of economic benefits for state and local governments, including new tax revenues, property taxes, increasing private sector spending, jobs and royalty payments to property owners (Kleinhenz and Smith 2011; Louis 2012). Yet, determining precise amounts of economic impacts has led to considerable debate between estimates/projections and actual results. Tables 5.13 and 5.14 are reflective of much of the early enthusiasm associated with Ohio's oil and gas industry.

Despite early sanguinity and confidence, other reports show lower tax collections and economic growth. Currently, the state receives 10 cents per barrel of oil and 2.5 cents per 1,000 cubic feet of natural gas (Pelzer 2015). Amounts are shown in Table 5.15. Revenue streams also support the ODNR's Oil and Gas Fund, which pays for costs associated with industry oversight and regulation.

Table 5.13 Early Projected Statewide Economic Impacts (from 2011 estimates)

	FY 2011	*FY 2012*	*FY 2013*	*FY 2014*
Value added	$162,030,036	$878,982,133	$2,980,378,198	$4,857,632,095
Labor income	$99,758,497	$571,543,463	$1,994,216,405	$3,298,757,195
Economic output	$291,574,770	$1,667,574,417	$5,823,268,396	$9,642,544,988
State and local taxes	$16,522,865	$73,422,148	$271,539,607	$433,528,922

Source: Kleinhenz and Smith (2011)

Table 5.14 Early Projected Severance Tax Collections

Year	*Severance Taxes*
2012	$434,862
2013	$2,967,123
2014	$15,080,854
2015	$32,368,301

Source: Kleinhenz and Smith (2011)

Table 5.15 Oil, Gas, and Sand Severance Taxes

Natural Resource	*Tax Rate*	*2010*	*2011*	*2012*	*2013*	*2014*
Natural gas	2.5 cents per 1,000 cubic feet	2,067,986	2,108,546	2,049,023	2,270,199	3,320,113
Oil	10 cents per barrel	487,165	474,886	467,111	567,887	652,574
Sand	2 cents per ton	293,012	330,431	329,857	368,858	302,094

Source: Ohio Department of Taxation (N.d.)

Employment Benefits

Resource extraction, transportation, and refining operations support employment opportunities across the state. Again, however, estimates can produce dramatically different numbers. According to estimates by Kleinhenz and Associates in 2011, for example, the shale oil and gas boom could support up to 105,000 jobs by 2014 and 170,000 by 2015 (if the region was fully developed). A forecast is shown in Table 5.16.

According to the Ohio Department of Jobs and Family Services, jobs related to extraction, drilling, support, pipeline construction, and transportation reported a significant increase between 2011 and 2014 (McParland 2014). Federal data also show job growth but not to the same extent as early estimations. According to the Bureau of Economic Analysis data, the number of full- and part-time employees directly involved in extraction has topped 20,000 since 2011–2012 (as shown in Figure 5.4). It should be noted that this number does not include jobs that are indirectly created through oil and gas production, which are included in the analysis performed by Kleinhenz and Associates.

Table 5.16 Projected Employment Impacts

	Direct Employment Impacts				
	2011	*2012*	*2013*	*2014*	*2015*
Mining and extractive	494	2,922	14,222	71,297	30,900
	Indirect Employment Impacts				
Support activities for mining	2,473	13,521	63,118	105,709	117,204
Retail trade	166	1,007	4,948	8,990	10,743
Professional and technical services	149	885	4,299	7,675	8,988
Administrative and support services	107	625	3,023	5,365	6,236
Ambulatory health care services	106	634	3,215	5,911	7,060
Construction	98	660	3,235	6,673	9,077
Food services and drinking places	71	434	2,156	3,994	4,940
Wholesale trade	54	321	1,539	2,722	3,162
Real estate	43	259	1,287	2,307	2,670
Personal and laundry services	33	201	1,010	1,834	2,158
Totals	**3,300**	**18,547**	**87,830**	**105,709**	**172,238**

Source: Kleinhenz and Smith (2011)

Table 5.17 The Oil and Gas Jobs Record of Ohio

Title	*2011 Q2*		*2015 Q2*		*Change*	
	Establishments	*Employees*	*Establishments*	*Employees*	*Establishments*	*Employees*
Crude petroleum and natural gas extraction	194	2,863	191	1,601	–3	–1,262
Natural gas liquid extraction	3	24	20	465	17	441
Drilling oil and gas wells	79	525	108	1,436	29	911
Support activities for oil and gas operations	181	1,292	309	4,252	128	2,960
Oil and gas pipeline construction	98	1,862	166	5,321	68	3,459
Pipeline transportation of natural gas	36	307	43	398	7	91
Core Industry Totals	591	6,873	837	13,473	246	6,600
Ancillary Industry Totals	12,864	166,913	12,961	180,282	97	13,369
Totals	**13,455**	**173,786**	**13,798**	**193,755**	**343**	**19,969**

Source: http://ohiolmi.com/OhioShale/Ohio%20Shale%20Report_2Q_2015.pdf

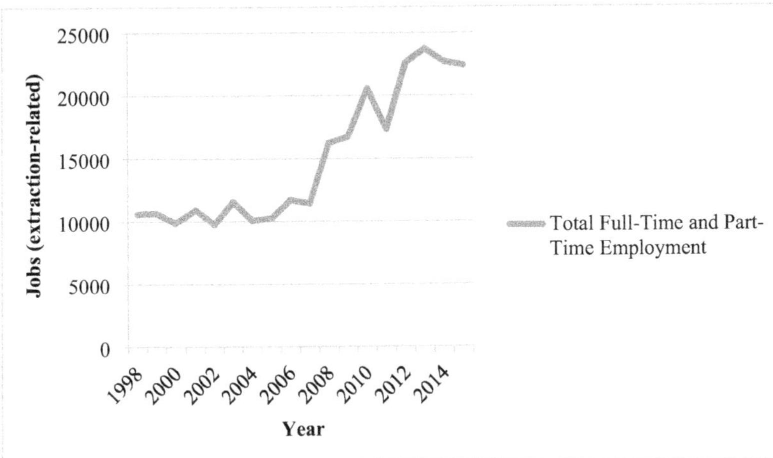

Figure 5.4 Ohio Oil and Gas Jobs (extraction-related)
Source: BEA (N.d.)

The average oil- and gas-related job in Ohio pays considerably more than the state average of $47,000. According to the Ohio Department of Job and Family Services (2016):

- Jobs in core shale-related industries between Q3 2014 and Q2 2015 averaged $75,071
- Jobs in ancillary shale-related industries between Q3 2014 and Q2 2015 averaged $63,147

Ohio's oil and gas operations have not been immune to fluctuations in oil and gas prices and industry volatility. Between 2014 and 2016, operators idled more than a third (24 of 59) of the rigs (equipment needed to drill) located atop the Utica Shale (Hoover 2015; Shingler 2015). This drop corresponds with an announcement from ODNR officials that the agency had issued fewer permits in 2015 (between 550 and 600) compared with 2014 (702) (Hoover 2015a). Production slowdowns have also led to workforce reductions:

- In September 2015, Chesapeake Energy Corp. announced that it had reduced its Ohio workforce by 15 percent.

(Knox 2015)

Support industries, such as Vallourec Star, which supplies steel tubing for drillers, have also been impacted by the industry's slowdown (nationally, since 2012 for gas producers and 2014 for oil producers).

- Early in 2015, Vallourec Star announced that it stopped production for three weeks. The decision followed reductions in its production schedules, changes to many of its supply contracts, and decreasing use of outside contractors.

(Shingler 2015a)

- TMK IPSCO, a pipe threading manufacturer used by the oil industry, announced that it had laid off 10 percent of its employees, including personnel at the company's Brookfield plant.

(WFMJ N.d.)

The boom and bust cycle has challenged Ohio's communities (Gearino 2015). Guy Coviello, the vice president of Government Affairs for the Youngstown/Warren Regional Chamber of Commerce, in reference to Vallourec's announcement, reflected, "You can't predict the cycles . . . but even though we knew it [boom and bust business], I think we all kept that knowledge in a closet somewhere" (Coviello as quoted by Shingler 2015a). Shawn Bennett, the Executive Vice President of the Ohio Oil and Gas Association, described the industry's economic situation in 2015 as "People are packing up and leaving and there are jobs being lost. Things just aren't going gangbusters the way they were a year ago" (Bennett as quoted by Shingler 2015a).

Changes in oil and gas employment are trickling down to local governments. During the summer of 2015, Carroll County Economic Development Director Glenn Enslen noted that only two drilling rigs were left in the county but that job losses for local residents were minimal. The downturn, according to Enslen, while not desirable, will give local governments, pipeline companies, and other processing firms in Carroll County an opportunity to "play catch up after being far behind the pace of production since drilling in the Utica began" (Enslen as quoted by Shingler 2015).

Costs

Similar to the national debate, Ohio's fracking intergovernmental battles involve the normative issue of whether oil and gas production near communities should be allowed and how it should be regulated. Many of these issues are summarized in Chapter 1. For Ohio, there are also place-specific factors that may influence local defiance. In this sense, the decision to defy the state is a function of Ohio's unique geography and a set of parochial concerns related to air quality, earthquakes, and drinking water safety (Henry 2013a). Anti-fracking groups have also argued that state authorities have failed to adequately respond to and meaningfully investigate citizen complaints related to oil and gas contamination (Johnson 2014). Brian Kunkemoeller, with the Ohio Sierra Club (an environmental interest group) stated, "We want the EPA to investigate all these complaints because we don't trust the substantiation by state agencies that these contaminations are not legit" (Kunkemoeller quoted in Knox 2014).

Air and Water Impacts

Opponents argue that production via fracking may exacerbates poor air quality by emitting air particulates and other ozone-causing chemicals. According to the Ohio Environmental Council (2014), air quality is harmed:

- During well construction, when operators may emit excess dust and other air-quality contaminates
- In operational stages, because heavy machinery, compressors, and other equipment consume up to 29,000 gallons of diesel fuel and other fracking fluids in as little as five days—which also leads to harmful emissions, including benzene, smog, and formaldehyde
- During operations, when compressors and other equipment emit pollution

Other research has confirmed that oil and gas production emits air pollutants, although at levels below EPA-established public health thresholds (Paulik et al. 2016).

Skeptics have also raised concerns over the impact of oil and gas on nearby surface waters. Again, evidence for fracking's impact on surface waters is mixed. It is clear, however, that the ODNR has received complaints and that activists firmly believe that fracking threatens drinking water.

Table 5.18 Surface Water Quality Impacts

Year	*Total Surface Complaints*	*Affected Water Systems*
2010	37	0
2011	54	2
2012	59	2
2013	33 (as of October 2013)	2

Source: Downing (2013), Knox (2014)

However, scientific studies have shown a complicated relationship between contamination and oil and gas development:

- A recent study of possible drinking water wells in Carroll, Columbiana, Stark, Harrison, and Belmont Counties found no evidence that nearby operators utilizing fracking had contaminated drinking water sources or that it had caused methane migration into drinking water.

  (Townsend-Small N.d.)

- Kell (2011) calculated that over a 25-year period, Ohio regulators detected approximately 185 cases of groundwater contamination caused by wastewater pit failure or poor well integrity, for an incident rate of approximately 0.1 percent (most of these incidents had not been related to horizontal drilling or fracking).
- Osborne et al. (2014) conducted tests for drinking water wells in portions of the Utica Shale underneath Pennsylvania and New York. They found methane concentrations in more than 80 percent (51/60) of wells across the included regions. They noted that "methane concentrations were 17 times higher on average in shallow wells from active drilling and extraction areas than in wells from non-active areas."

Cumulative Environmental Impacts

Table 5.19 presents the cumulative environmental impacts as measured by Environment America, an anti-fracking advocacy group.

The ODNR maintains that fracking, when property regulated, is environmentally safe (Smith 2015b). It added that between 2010 and October

Table 5.19 Aggregated Environmental Impacts of Fracking

Environmental Harm	*2012 Impact*
Acres damaged since 2005	1,600
Greenhouse gas emissions (metric tons of carbon dioxide-equivalent) (between 2005–2012)	420,000 tons
Particulate matter	100 tons
Nitrogen oxide	1,700 tons
Carbon monoxide	2,600 tons
Volatile organic compounds	200 tons
Sulfur dioxide	6 tons

Source: Ridlington and Rumpler (2013)

2013, the Division of Oil and Gas Resources Management investigated more than 180 natural gas well/site complaints. It found that most complaints were related to older vertical wells (and not horizontally drilled wells) and that spills rarely impacted surface water quality (Downing 2013a).

Injection Wells

Ohio operators may clean, store, dispose of, or recycle wastes generated by oil and gas operations (Main 2013). Mark Bruce, an ODNR spokesman, noted that wastes are treated the same as other industrial byproducts. Through inspections, rulemaking, and oversight, the ODNR ensures that companies follow state and federal laws (Arenschield 2014a; Davis and Fisk 2017).

Concerns over injection well siting (Class II oil and gas disposal wells) have also led to intergovernmental tensions in Ohio:

- The city of Wooster has battled Enviro-Tank Clean over the location of a wastewater treatment plant. The city argues that the plant emits noxious odors and that Enviro-Tank Clean never informed city officials that it would be receiving and processing fracking wastewaters. Richard Benson, Wooster's law director, worried about "what sort of protocols do they have for safety?" (Benson as quoted by Arenschield 2014a), *Wooster v. Enviro-Tank Clean, Inc.*, 2015-Ohio-1876.
- Voters in the cities of Cincinnati, Yellow Springs, and Niles banned injection wells in 2012.

  (Marshall 2012a)

- Voters in the city of Mansfield approved a home rule charter amendment that enables the city to ban injection wells.

  (Marshall 2012a)

A number of county officials in Ashtabula, Athens, Portage, and Trumbull Counties have criticized the state's regulatory framework addressing oil and gas wastes. Casey Kozlowski, a county commissioner in Ashtabula County, noted the inequity of out-of-state wastes arriving in Ohio, "They're [Pennsylvania] seeing the production; we're seeing the waste" (Kozlowski as quoted by Dennis 2015a). Mike Fitchet, Ashtabula County's Director of Emergency Management, reported that local first responders needed more information relative to well sitemaps, routes in and out of the site, and the type/nature of chemicals stored at the site. Ashtabula County Auditor Roger Corlett noted, "Property values are influenced by property sales, and though sales continue around well sites . . . I think in the long run, you'll see a great decline in value" (Corlett as quoted by Dennis 2015a).

Trumbull Township Trustee Ron Tamburrino also offered a number of insights into ways that operators could work with residents and other local governments to assuage their concerns (Dennis 2015a):

- Committing to greater public involvement during permit applications and improved notification of stakeholders (local governments, landowners, and schools)
- Holding hearings for well permit objections
- Expanding environmental review areas and the introduction of traffic impact studies
- Lengthening public comment periods
- Adding site aesthetic standards and activity restrictions

Induced Seismicity

Ohio's experiences with induced seismicity have also complicated the intergovernmental management of oil and gas production. The state's history does involve some minor earthquakes along small but known fault lines. Earthquakes began increasing in frequency in 2011. Between March and December 2011, for example, Ohio seismologists recorded more than a dozen earthquake events near Youngstown, with one quake measuring 4.0 in magnitude (strong enough for residents to feel it). Early in 2012, Governor Kasich ordered a moratorium on wastewater disposal wells within a seven-mile radius of the quake's epicenter (Henry 2013, 2013a; Ohio EPA 2014b).

During March 2014, seismologists recorded 77 small seismic events in Mahoning County. The earthquakes, located near an unknown fault line, triggered the ODNR to suspend nearby drilling (Downing 2015). Following the spike in quakes, the ODNR issued new rules that required wastewater operators to provide more data to regulators and to be subject to a greater number of random inspections (Brown 2015).

The presence of man-made earthquakes appears to be a significant source of concern for local leaders. Following the 2011 cluster of earthquakes near Youngstown, Youngstown city leaders, including the mayor (and State Representative Robert Hagan), supported a moratorium on the use of injection wells in the area. While supporting injection well suspensions, Governor Kasich argued that the earthquakes are isolated events and that the city's call for a moratorium would jeopardize economic recovery in Eastern Ohio and threaten jobs (Niquette 2012a).

Summary on Ohio's Intergovernmental Issues

The State Supreme Court, in its *Munroe Falls* decision, affirmed Ohio's state-centric approach to managing oil and gas production. Following the *Munroe Falls* decision, however, Beck Energy requested that the Ohio

Supreme Court prevent the city of Munroe Falls from using its general zoning powers to preclude development (rather than a specific ordinance as was the case in the earlier *State ex rel. Morrison v. Beck Energy Corp.*, 143 Ohio St.3d 271 (2015). The court dismissed Beck's request in November 2015, leaving unanswered the question of whether cities can use their traditional zoning authority to impede, regulate, or restrict oil and gas development as part of their general zoning authority (Smith 2015). Despite strong state-level support for continued development in the Utica Shale, intergovernmental questions such as to what extent can cities/townships limit development through their traditional zoning authority, what is the ideal severance tax rate, should the state and to what degree should the state distribute revenues to impacted local government, and how to resolve local concerns related to injection wells are still sorting themselves out. The answers to these questions require new and more effective forms of governance that build upon the strengths of the ODNR, OEPA, and Ohio's cities and townships.

6 Oil and Gas in Texas

Intergovernmental challenges in Texas are related to the state's long-standing tradition of local control and to its century-plus history as a leading oil- and gas-producing state. Thus, from the perspective of Texas it is important to identify the notable state laws and rules, and how state courts have weighed previous disputes related to oil and gas. Like the two previous states, this chapter is also divided into two sections. The first summarizes the legal frameworks that govern intergovernmental policymaking and production. It is guided by questions such as how does the Texas Railroad Commission (RRC) manage/oversee oil and gas production? What specific regulations address drilling and fracking? Equally important, what roles do local governments play when it comes to oil and gas production? How have state lawmakers reacted to local defiance in Denton, Texas? How do state courts approach intergovernmental cases that have both state and local impacts? Have activists looked to new venues to bring about policy change? The answers to these questions form half of the local defiance equation (i.e., the state policies that local governments are challenging).

Particular regions, counties, or communities uniquely experience the costs and opportunities related to oil and natural gas production. In Texas, production in the Barnett, Permian, and Eagle Ford regions has placed additional governing challenges on the doorsteps of city hall. Given this, part of unpacking local defiance is an examination of the political 'geology' of oil and gas within Texas. In other words, how has development impacted the state and its local governments in terms of revenues, jobs, and environmental harms? Are local governments benefiting or are they experiencing an unfair share of the costs associated with development? These answers add into intergovernmental discussions focused on developing new and inventive forms of state-local governance.

Following is a selected set of stakeholder reactions to Texas HB 40 (2015). The law followed the city of Denton's voter-passed ban on fracking and included new restrictions on the ability of the city to regulate oil and gas operations:

- Governor Greg Abbott described HB 40 as legislation that protects private property rights from overzealous local governments; "those who own their own property will not have the heavy hand of local government deprive them of their rights."

  (Abbott as quoted by Baker 2015b)

- Governor Abbott also observed that HB 40 "avoids a patchwork quilt of regulations that differ from region to region, differ from county to county or city to city" (Abbott as quoted by the Guardian Staff 2015). He added that a uniform set of state regulations would reduce industry's costs and improve its operational efficiencies.
- Bill Lane, the city of Mansfield's Public Safety Director and the Senior Staff Attorney, criticized HB 40's 'commercially reasonable' language as being overly vague: "The problem is, what the hell is 'commercially reasonable'?" (Lane as quoted by Malewitz 2015). As part of a compromise between industry, the Texas Municipal League, and stakeholders, HB 40 permitted local governments to pass commercially reasonable oil and gas regulations.
- City of Denton Mayor Chris Watts, while acknowledging that HB 40 made Denton's fracking ban unenforceable reflected, "I think anytime that a vote of a community is nullified by some action that is disappointing . . . [the city-wide ban was] based on what we needed for the city."

  (Watts as quoted by Baker 2015)

Why Texas?

The intergovernmental story of Texas relative to oil and gas production reveals a number of important lessons for stakeholders. First, Texas has a long history of encouraging oil and natural gas production alongside a rich tradition of protecting home rule authority. Texas's local governments used their authority to establish a wide variety of setback distances (including some that industry described as de facto bans), site security/aesthetics, venting/flaring restrictions, inspections, and other surface-level requirements. Recent legislation has reduced this authority. However, Texas cities may still influence the scale, character, and location of production, with a state-set standard of commercial reasonableness. Second, resource production within the Dallas/Fort Worth region, in areas near San Antonio, and West Texas represent new challenges to cities. In 2013, approximately seven million Texans lived or had lived within one mile of a well (not necessarily a well that was or will be fracked) (Gold and McGinty 2013). Finally, oil and gas development represent a large part of the state's identity and economy, but examples of local defiance are bubbling up. During

2014–2015, one city in the Dallas-Fort Worth Metroplex banned fracking while several other communities imposed large setbacks and other restrictions on development. Yet, a majority of communities atop the Barnett Shale and many others across the Permian Basin and the Eagle Ford Shale have taken steps to encourage additional drilling/development.

Similar to Colorado and Ohio, the Texas experience with oil and gas production enriches the study of natural resource federalism and offers lessons to other communities and states. Beyond the reasons listed in the preceding paragraph, Texas's state-local conflicts are a function of both state and local actors. Thus, by understanding from an intergovernmental perspective what has worked and what has not, best practices may be distilled that can be applied within Texas and by jurisdictions located elsewhere.

Texas's Policymaking Context

State constitutional provisions, statutes, regulations, and property rights have encouraged oil and gas production since the early 20th century (Davis 2012; Rahm 2011). This support has continued throughout the shale oil and gas revolution. Since 2008, state lawmakers, including those with the Railroad Commission of Texas (RRC) and the Texas Commission on Environmental Quality (TCEQ) have promulgated industry-supported chemical disclosure rules, air and water quality standards, severance tax incentives, seismicity, well design, placement, and spacing requirements (Davis 2012). State lawmakers also enacted HB 40, an industry-backed law that limited local regulations to a commercially reasonable standard, among other restrictions. (Davis 2012; Price 2012). Oil and gas production also produces sizeable economic benefits (employment and revenues) that further incentivize production throughout the state's shale regions.

Constitutional and Legal Provisions

A number of laws in Texas encourage robust oil and gas production, including the state's property laws, industry incentives, water regulations, and other environmental laws.

Property laws in Texas generally encourage oil and gas development. Maxwell (2009, 356–357) identified several examples, as noted in Table 6.1.

Other major laws that can impact the oil and natural gas industry are given in Table 6.2.

Since 2005, the state has also enacted a number of financial incentives to increase production from oil and gas wells. A summary of

Table 6.1 The Mineral Estate in Texas

Law	*Impact on Oil and Gas Development*
Texas splits property ownership between the surface and mineral estate, with the latter being dominant.	The mineral estate is entitled to the reasonable use of the surface in order to extract the mineral (Rahm 2011).
State law permits the mineral estate to be held in perpetuity.	The mineral owner maintains his or her right to develop the estate with no time restrictions, limitations, or requirements relative to development, *Arco v. Railroad Commission*, 346 S.W.2d 801 (Tex. 1961); *Coastal Oil & Gas Corp. v. Garza*, 268 S.W.3d 1 (Tex. 2005).
The mineral owner may extract as much of the mineral that will flow from of the well, subject to the RRC's conservation standards, *Arco v. Railroad Commission*, 346 S.W.2d 801 (Tex. 1961), *Coastal Oil & Gas Corp. v. Garza*, 268 S.W.3d 1 (Tex. 2005).	Operators are encouraged to produce as much of the mineral as possible, subject to state conservation mandates, market dynamics, and available infrastructure.

Table 6.2 Other Environmental and Energy Laws

Name and Year	*Policy Area*	*Agencies*
Texas Pollution Control Act (1961)	Waste prevention and reduction	Texas Commission on Environmental Quality
Texas Water Quality Act (1967)	Water quality	Texas Commission on Environmental Quality and Railroad Commission
Texas Injection Well Act (1961)	Injection wells (other than from the oil and gas industry)	Texas Commission on Environmental Quality and Railroad Commission
Texas Solid Waste Act (1969)	Hazardous waste	Texas Commission on Environmental Quality
Texas Clean Air Act (1965)	Stationary and mobile air sources	Texas Commission on Environmental Quality
Renewable Generation Requirement (1999)	Renewable energy— 5,880 MW by 2015 10,000 MW by 2025 (goal)	

Source: Texas Commission on Environmental Quality (2014)

severance tax incentives is available here: www.rrc.state.tx.us/oil-gas/publications-and-notices/texas-severance-tax-incentives-past-and-present/presenttax/. Many operators that utilize fracking have also taken advantage of the state's 2003 high-cost drilling tax break. According to a Legislative Budget Board (LBB) report, the tax break defined "late 1970s production factors" as the baseline for determining 'high-cost drilling' operations. However, 'high-cost drilling' of the 1970s is what constitutes many of today's fracking operations. According to the board, "in 2009 over a third of the wells approved as 'high cost' in the state (2,128 of 5,967 wells) paid an effective tax rate of zero" (Fehling 2012b).

Local Powers

Under Texas law, cities are considered as a general law or home rule community. Cities with more than 5,000 residents may adopt a local charter and become a home rule community. Once adopted, home rule cities possess the "full power of local self-government," including the power to adopt municipal ordinances, issue planning and zoning plans and author transportation plans, and enact other regulations aimed at promoting public health and safety (Welch 2007, 144–146) (McFarland 2013). By comparison, communities with populations below 5,000 are classified as general-law cities type A, B, and C (based on population). Texas law permits general-law cities to establish zoning standards (LGC 211, 213), create and maintain police and fire services (LGC 341–342), construct and maintain streets (LGC 311–312), and abate nuisances (LGC 217) (Mueller 2015).

In Texas, municipal oil and gas regulation dates back to the 1970s. Today, many Texas cities regulate some aspect of oil and gas production, including:

- Comprehensive regulatory standards, which mandated that operators receive a local permit before drilling
- A requirement that operators allow municipal officials or their authorized designees to conduct inspections
- Policies aimed at mitigating the impacts of well flaring and venting, site security, and well pad aesthetics
- The issuance of fines for non-compliance

Portions of the city of DISH's Texas oil and gas ordinances are included in Table 8.5.

The city has also banned oil and gas drilling in its floodplains, in areas adjacent to floodways, and in zoned areas that do not specifically permit development (DISH 2014).

Table 6.3 DISH, Texas, Urban Drilling Surface Regulations

Policy Area—Surface Regulation	*Example of Municipal (DISH, Texas) Requirements*
Notification	The operator must post a sign at the well's entrance at least 48 hours before a fracking operation begins.
Fluid recovery	The operator may only recover fluids during daylight hours unless the local inspector authorizes non-daylight recovery hours.
Security	The operator must post a watchman during operations.
Venting	Venting directly into the atmosphere is not allowed and operators must first direct flaring and any flow through separation equipment or into a portable tank.
Noise	Operations cannot exceed the ambient noise level of 10 decibels. This level is lowered to five decibels for backflow operations during nighttime hours.
Odors	Fracking operations must not emit odors that are "extremely repulsive to the physical senses of ordinary persons which annoy, discomfort, injure, or inconvenience the health of any appreciable number of persons."
Inspections	Personnel may enforce the city's provisions and conduct inspections to ensure compliance.
Permits	Operator must apply for and obtain a gas well permit for the drilling, re-drilling, deepening, re-entering, activating, or converting of each well.
Moratorium	City lawmakers temporarily suspended the issuance of oil and gas permits. The 90-day moratorium applied to all permits and applications needed to drill within the city.
Setbacks	Within 200 feet to any fresh water well. The measurement shall be in a direct line from the closest well bore to the fresh water well bore; or
	Within 1000 feet from any building used, or designed and intended to be used, for human occupancy.
Restrictions are limited to zoned areas within the city that permit drilling.	

Source: DISH (2014, 2014a, 2014b)

House Bill 40

In 2015, following the citizen-backed ballot initiative that banned fracking within Denton, Texas, legislators enacted HB 40 (Baker 2015, 2015a). The law preempted local authority to ban fracking among other restrictions on local policymaking authority:

- Rules cannot regulate subsurface activity.
- Rules (including setbacks) must be "commercially reasonable" (meaning "a condition that would allow a reasonably prudent operator to

fully, effectively, and economically exploit, develop, produce, process, and transport oil and gas, as determined based on the objective standard of a reasonably prudent operator and not on an individualized assessment of an actual operator's capacity to act").
- Rules must not effectively block or preclude oil and gas operations from occurring within the city.
- Rules may be preempted by other state or federal laws.

(Heinkel-Wolfe 2015)

- Rules presume commercial reasonableness for ordinances that have allowed oil and gas activities for at least after five years (beginning in 2015).

Under HB 40, cities still regulate emergency response, including fire and EMT functions, and oil and gas truck routes; assess road quality/impact fees; limit industrial noise; establish setbacks (from homes, schools, and businesses; must be commercially reasonable); and restrict lights from oil and gas sites (Sakelaris 2015).

State Regulations

The RRC and the TCEQ are the state's two main regulatory agencies charged with overseeing the state's oil and gas resources, with the RRC being the predominant institutional actor (Price 2012).

The legislature charged the TCEQ to protect the state's environment in a manner consistent with the principles of sustainable economic development.[1] Before drilling, operators require authorization with permit/license (the type depends on the type of facility and contaminants). The agency also establishes controls and limits for industry's particulates, volatile organic compounds (VOCs), nitrogen compounds, and sulfur (TCEQ 2016) emissions/pollutants. Regulators at the TCEQ also manage surface water diversions but not stormwater discharges (RRC 2014; TCEQ 2014c; Texas Alliance of Groundwater Districts 2013).

The RRC's regulatory role stretches back to 1891, when voters authorized the state legislature to establish the Texas Railroad Commission (RRC). At the time, the agency's jurisdiction was limited to the state's railroads. Under the 1917 Pipeline Petroleum Law (SB 68), state lawmakers broadened its mandate to include gas production and intra-state oil and gas distribution pipelines. The 1919 Oil and Gas Conservation Law (SB 350) further expanded the RRC's role in oil and gas management. Shortly thereafter, the RRC issued its first set of standards, addressing spacing, safety, and statewide energy conservation (Riley 2007; University of Texas Libraries 2014). Today, the agency works to ensure efficient production of the state's mineral resources, protect private property, and preserve public safety (University of Texas Libraries 2014).[2] Despite no formal requirements that commissioners have an industry background,

Davis (2012) described the agency's leadership team as generally inclined to be supportive of oil and natural gas interests.

RRC permitting rules follow the general process outlined in Chapter 1.[3] Other regulations are noted in Table 6.4.

Regulatory power, subject to legislative changes (such as HB 40), is shared among local governments, the RRC, and the TCEQ, with a summary presented in Table 6.5.

Table 6.4 Other Texas Fracking Laws/Regulations

Regulatory Area	*Law/Regulation*
Pre-drilling water well testing	No evidence of regulation in effect
Water withdrawal restrictions	Addressed in permit
Casing and cementing depth requirements	Performance standard
Venting and flaring	Notification and approval required
Fluid storage	Pits allowed and regulated for all fluids, including freeboard and liner requirements and five-year tracking requirements
Underground fluid injection	Allowed in areas that are not producing oil, gas, or geothermal resources. Fluid injection areas must be separated by water formations by impervious zones that offer adequate protection against contamination (RRC 2014a)
HB 3328 (Texas Hydraulic Fracturing Fluids Disclosure Act)	Requires that operators provide FracFocus with the specific fluids used in the fracturing treatments
Seismicity	Requires that operators provide the RRC with additional United States Geological Survey information: • Seismic data for the previous 100 years for ten square miles around the proposed drill site. • Additional information could be required if the proposed site is in an area that may be at a heightened risk for a seismic event (could include logs, geological cross-sections, and pressure front boundary calculations). • Industry may be required to conduct additional monitoring and reporting of Class II disposal well pressures, volumes, and rates. Under the new rules, the RRC may modify, suspend, or terminate disposal well permits should they be linked to induced seismicity (RRC 2014, 2014a).

Source: Richardson et al. 2013

Table 6.5 Major Regulatory Actors and Policy Areas

Focal Area	*Reporting Agency*
Odor complaints	TCEQ
Noise complaints	Local law enforcement
Dust (from public roads)	Local government
Traffic complaints	Local law enforcement
Private wells (drinking water)	RRC
Public water supply (drinking water)	TCEQ
Spills associated with the exploration and production of oil and gas	RRC
Wastes from oil and gas (including injection wells)	RRC
Air emissions during drilling or exploration activities	RRC
Wastes produced during the transportation of crude oil and gas by rail, truck, barge, or tanker and refined petroleum products via pipeline	TCEQ
Wastes produced at oil field service facilities that provide equipment, materials, or services to the oil and gas industry	TCEQ

Source: TCEQ (2015)

Judicial Decisions About Local Control and Oil and Gas

Local oil and gas regulations are presumed valid if they satisfy a three-part analytical framework as follows (*Klepak v. Humble Oil & Ref. Co.*, 177 S.W.2d 215; *Unger v. State*, 629 S.W.2d 811 [Tex.App.-Ft. Worth 1982]):

1. The ordinance must not address policy areas that the state has preempted with unmistakable clarity.
2. The ordinance must be validly enacted and its language must not be arbitrary and or unreasonable.
3. The ordinance must advance a legitimate interest and must be reasonably related to protecting public safety and general welfare.

(McFarland 2013)

Texas state courts have affirmed local oil and natural gas regulation for more than 50 years. In 1944, with its *Klepak v. Humble Oil & Refining Co.* (177 S.W. 2d 215), the First Court of Appeals upheld a local ordinance that limited the number of wells to one per drilling block. The decision reasoned that unless city actions are facially unreasonable, arbitrary, or discriminatory they should be presumed valid as a part of the city's police powers. The *Klepak* jurisprudence shaped subsequent oil and gas intergovernmental questions through the early 1980s.

In 1982, the city of Burkburnett, Texas, fined a private property owner (Unger) for drilling within its limits before receiving a city-drilling permit. Unger claimed that the city's permitting process was invalid because it went beyond protecting public health to proscribing technical oil and gas procedures (*Unger v. State*, 629 S.W.2d 811 [Tex. App. 1982]). The Fort Worth Court of Appeals ruled in favor of the city, noting that unless the municipal ordinance is facially unreasonable or is a regulatory taking, it should be presumed to be a valid exercise of local land use power (Rutherford 2014).

The outcomes mirrored each other, but the courts' differing logics in *Klepak* and *Unger* are worth noting:

- In *Klepak*, the court considered the local ordinance to be part of the city's police powers.
- The court in *Unger*, however, treated the city of Burkburnett's ordinance as part of its land use and zoning powers.

Riley (2007, 371) summarized the importance of *Unger* as opening "a new line of reasoning supporting municipal authority regulating oil and gas development in a manner similar to other forms of land use restrictions" (see also McFarland 2013). Subsequent cases have followed

Table 6.6 Intergovernmental Oil and Gas Cases and State Courts

Court Case	*Decision*
Shelby Operating Co. v. City of Waskom, 964 S.W. 2d 75 (Tex. App. 6th Dist. 1997)	The case began in 1947 when Shelby acquired a 303-acre mineral lease for an area located outside the city's 1945 limits. In 1981, the city annexed portions of surface estate that were located above the mineral estate. The city then sold the surface estate to the Aztec Manufacturing-Waskom Partnership in 1986. A year later the city enacted a 500-foot setback requirement applicable to all structures unless the surface owner provides consent. In 1996, Shelby applied for a local permit to drill a well within the city's 500-foot setback zone. Aztec did not give its consent, and the city rejected Shelby's permit request. The court found in favor of the city, noting that the ordinance was related to its home rule powers to regulate land use and protect public safety (Riley 2007).
Midstream Gas Servs., LLC v. City of Grand Prairie, No. CIV.A.3:08CV1724-D, 2008 WL 5000038	The court denied a preliminary injunction against a city ordinance that required operators to install landscaping and to follow noise limits. It, however, found that state rules preempted portions of the ordinance that addressed pad fencing (Goho 2012).

Source: Author generated

a similar line of logic and have protected municipal regulation unless it violates the three-part test summarized previously in this chapter.

The Oil and Gas Debate in Texas

Intergovernmental politics in Texas resemble many of the challenges of other oil and gas states. Supporters argue that production has contributed and will continue to contribute to a strong and robust state economy. Through a standardized set of state policies, they argue, operators are better positioned to find operational efficiencies and to minimize costs. Opponents argue that the economic impacts of oil and gas are fleeting, and its costs to infrastructure and the environment are dangerously high and permanent.

Oil and Natural Gas Production

The Texas oil and natural gas 'story' dates back to early in the 20th century when Texas wildcatters discovered a 'gusher' at the Spindletop oil field in 1901. By the early 1920s, significant reserves were also discovered near Laredo, White Point, and Kingsville. That legacy continues, with the boom 'beginning' for gas around 2005–2006 and for oil around 2010. In 2012, operators produced more than 7.1 trillion cubic feet (Tcf) of natural gas. In 2015, that number increased to 8.3 Tcf. The state expects gas production to remain strong, despite some contraction in the industry (Davis 2012; RRC 2016; Pless 2010; Rahm 2011). Proven oil reserves have also ballooned since 2006. In 2006, the Energy Information Administration (EIA) estimated that the state held 4,871 million barrels of oil, and as of 2013, that estimate had nearly tripled to 11,759 million barrels of oil (EIA 2013a) (Tables 6.7 and 6.8).

Texas production has also benefited from a large and interstate network of oil and gas pipelines. West Texas lines supply consumers in Arizona, New Mexico, California, and several Midwestern markets.

Table 6.7 Recoverable Reserves in Texas (2014)

Resource	*Reserves*	*Percent of U.S. Total*
Crude oil	11,759 million barrels	36.4%
Dry natural gas	78,866 billion cubic feet	25.6%

In November 2016, the United States Geological Survey announced that the Wolfcamp Shale contains technically recoverable reserves estimated at 20 billion barrels of oil and 16 Tcf of natural gas (Hersher 2016).

Source: EIA (2016d)

Table 6.8 Locations of Reserves and Production

Producing Region	*Location*
Eagle Ford Shale	Southern Texas (near San Antonio)
Barnett Shale	North Texas (Dallas-Fort Worth region)
Permian Basin	West Texas
Wolfcamp Shale	West Texas

Source: EIA (2013a)

Table 6.9 Texas Outflow Pipeline Capacity—Million Cubic Feet/Day

State From	*State To*	*2009*	*2010*	*2011*	*2012*	*2013*	*2014*
Texas	Arkansas	2,072	2,072	2,184	2,184	2,184	2,184
	Louisiana	12,022	15,152	17,652	17,652	17,652	17,652
	Mexico	2,551	2,551	2,551	2,551	2,917	5,017
	New Mexico	4,637	4,637	4,637	4,637	4,637	4,637
	Oklahoma	5,964	5,964	5,964	5,964	5,964	5,829
Texas Total		**27,246**	**30,376**	**32,988**	**32,988**	**33,354**	**35,319**

Source: EIA (2016f)

Production from the Barnett and Eagle Ford shales has also supported the construction of additional pipeline capacity that feeds Southeast and Midwest markets, as shown in Table 6.9 (EIA Natural Gas 2008).

The state's access to the Gulf of Mexico has also incentivized oil and natural gas production via energy exporting (Shauk 2014). During the peak of the boom, Cheniere Energy, for example, announced plans to build a second gas liquefaction terminal, allowing it to be shipped to customers across the globe:

- Cheniere has signed a 20-year, $5.5 billion agreement with the British utility, Centrica, to supply enough gas to heat 1.8 million homes. (Weber 2013)

- It has also entered into a 20-year, $5.6 billion deal to export 400,000 metric tons of liquefied natural gas (LNG) to Iberdola (a Spanish firm) from a (planned) LNG terminal in Corpus Christi, Texas. (Cama 2014)

By 2020, Cheniere anticipates that producers will ship about 6 percent of all domestically produced natural gas to international customers. As of

October 2015, the Federal Energy Regulatory Commission was reviewing proposals for several other gas export terminals in Texas. Proponents estimate that, if fully developed, liquefied natural gas exporting would support more than 150,000 jobs by 2035 (Del Valle 2015).

The confluence of a favorable political climate, a sympathetic regulatory environment, and additional demands for Texas's hydrocarbons has led to a significant increase in oil and gas production in recent years, as demonstrated in Figures 6.1 through 6.3, with noticeable jumps in 2005 and 2010.

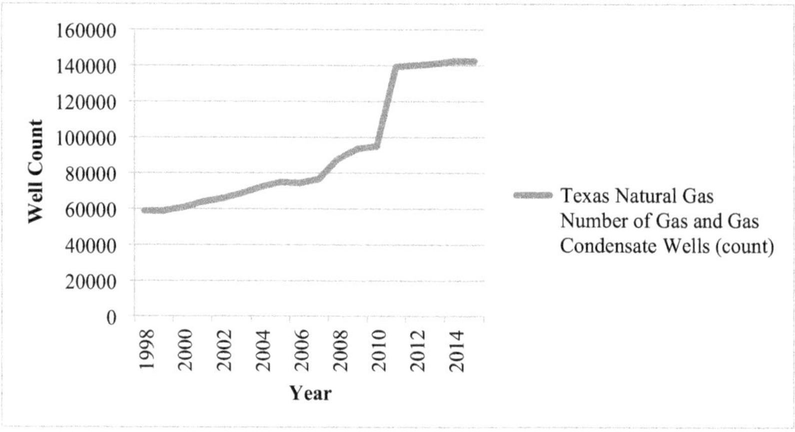

Figure 6.1 Texas Natural Gas Number of Gas and Gas Condensate Wells
Source: EIA (2014)

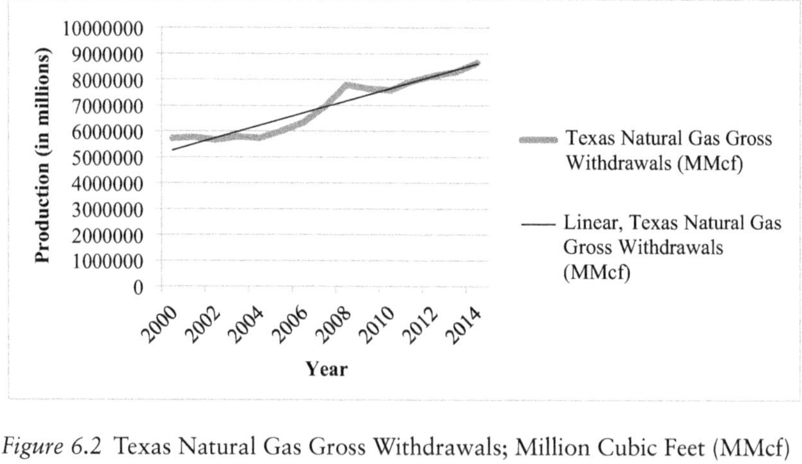

Figure 6.2 Texas Natural Gas Gross Withdrawals; Million Cubic Feet (MMcf)
Source: EIA (2014a, 2014b)

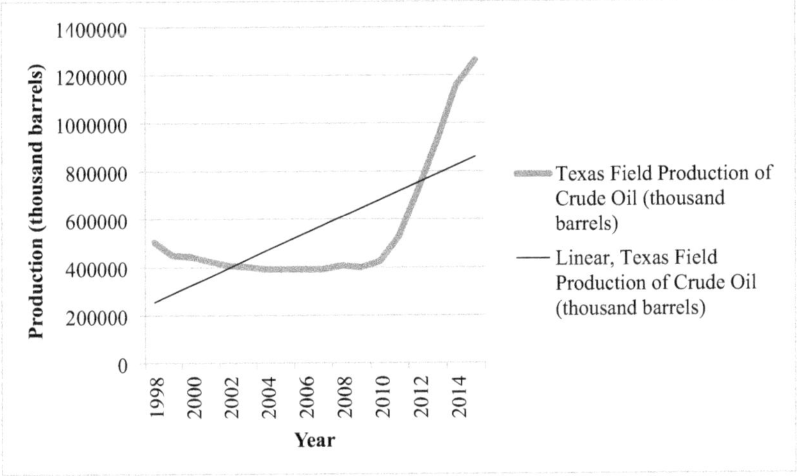

Figure 6.3 Texas Field Production of Crude Oil
Source: EIA (2014c)

Location of Development

Oil and gas production yields statewide benefits. It, however, is concentrated in a handful of major producing regions: Barnett Shale (North Texas), Eagle Ford Shale (South Texas), and the Permian Basin (Far West Texas), meaning that production's costs and benefits are clustered in specific counties, as shown in Table 6.10.

Table 6.10 April 2015—Top Ten Oil and Gas Counties, by Preliminary Production

County	*Crude Oil (barrels)*	*County*	*Total Gas (millions of cubic feet)*
1. Karnes	5,422,409	1. Tarrant	51,988,466
2. La Salle	4,485,166	2. Webb	31,777,105
3. De Witt	3,435,580	3. Johnson	28,170,963
4. Gonzales	3,277,904	4. Panola	22,555,548
5. McMullen	2,819,333	5. Wise	18,327,726
6. Andrews	2,724,419	6. De Witt	18,208,243
7. Upton	2,156,708	7. Dimmit	16,489,629
8. Ector	2,078,548	8. Denton	15,779,884
9. Martin	2,055,120	9. Karnes	15,717,876
10. Dimmit	2,008,103	10. Wheeler	14,164,240

Source: RRC (2014c)

State Economic Benefits

Oil and gas firms are significant contributors to the Texas economy. They directly employ thousands of Texans and indirectly support thousands of other jobs (Texas Oil and Gas Association [TXOGA] 2013, 2013a). Additionally, the taxes levied on the oil and gas industry generate billions of dollars for state and local governments. In fiscal year (FY) 2012, for example, oil and gas severance tax collections were approximately $3.6 billion. Based on the state's allocation formulas, primary schools in Texas received $905 million (25 percent) of that amount, with the remainder deposited into the state's general fund. The state also collected $9 million in fees and other assessments. Finally, the state also leases public lands for oil and gas extraction. In FY 2012, state lease revenues totaled $1.3 billion (Raimi and Newell 2014). When aggregated, the Texas oil and gas sector produces an estimated statewide economic impact of $290 billion (for FY 2015). This number jumps to more than $628 billion, if indirect economic contributions from oil and gas are included, as shown in Table 6.11.

The state levies taxes on oil and gas production and services that help fund its school and general revenue funds. It collects a gas production tax of 7.5 percent of the market value of natural gas (Table 6.12). The state also charges a severance tax on oil production, currently set at 4.6 percent of the market value of oil extracted within the state (Table 6.13). Once collected, 25 percent of each is reserved for the state's school fund, with the general revenue fund receiving the remainder (Texas House Committee on Business & Industry 2015).

Table 6.11 Oil and Gas Industry Estimated Contributions to State Economy (in billions)

Category	*Direct*	*Indirect*	*Total*
Crude petroleum and natural gas extraction	$168.8	$102.9	$271.7
Natural gas liquid			
Drilling oil and gas wells			
Support activities, oil/gas operations	23.8	35	58.8
Natural gas distribution	6.2	6.1	12.3
Petroleum refineries	56.3	113.6	169.9
Petrochemical manufacturing	11.6	48.1	59.6
Oil and gas field machinery and equipment	11.7	16.8	28.5
Petroleum merchant wholesalers	3.6	2.2	5.8
Pipeline transportation	8.7	13.4	22.1
Total	**290.6**	**338.1**	**628.7**

Source: Texas House Committee on Business & Industry (2015)

Table 6.12 Natural Gas Production Tax (in thousands)

Year	*Revenues*	*Year*	*Revenues*	*Year*	*Revenues*	*Year*	*Revenues*
2000	697,666	2004	1,392,436	2008	2,684,648	2012	1,534,630
2001	1,596,886	2005	1,657,086	2009	1,407,739	2013	1,495,203
2002	628,497	2006	2,339,147	2010	725,538	2014	1,899,582
2003	1,069,864	2007	1,895,488	2011	1,109,718		

Source: Texas Comptroller of Public Accounts (2014)

Table 6.13 Crude Oil Production/Severance Tax (in millions)

Year	*Revenues*	*Year*	*Revenues*	*Year*	*Revenues*	*Year*	*Revenues*
2000	415,800	2004	495,443	2008	1,436,243	2012	2,102,389
2001	441,922	2005	681,294	2009	883,774	2013	2,989,542
2002	337,857	2006	861,659	2010	1,008,074	2014	3,872,279
2003	422,879	2007	834,374	2011	1,472,111		

Source: Texas Comptroller of Public Accounts (2014)

The state also collects an oil regulation tax, set at 3/16 of 1 cent per barrel of crude oil—with all revenues going to the general fund (Table 6.14). Finally, the state also collects a small well service tax (Table 6.15).

Table 6.14 Oil Regulation Tax Revenue (in thousands)

Year	*Revenues*	*Year*	*Revenues*	*Year*	*Revenues*	*Year*	*Revenues*
2000	820	2004	668	2008	636	2012	879
2001	658	2005	597	2009	737	2013	1348
2002	804	2006	702	2010	590	2014	1792
2003	708	2007	651	2011	736		

Source: Texas Comptroller of Public Accounts (2014)

Table 6.15 Total Oil Well Service Tax Revenue (in thousands)

Year	*Revenues*	*Year*	*Revenues*	*Year*	*Revenues*	*Year*	*Revenues*
2000	7,359	2004	14,627	2008	49,749	2012	122,149
2001	12,156	2005	22,199	2009	36,114	2013	112,698
2002	13,196	2006	34,591	2010	26,650	2014	129,604
2003	11,899	2007	43,564	2011	79,381		

Source: Texas Comptroller of Public Accounts (2014)

Oil and gas production has also benefited from the state's rainy day fund. During FY 2010, state lawmakers shifted $450 million into the fund, pushing it over $8 billion. By November 2015, the fund contained $10.4 billion. Even before the widespread use of horizontal drilling and hydraulic fracking, the state's rainy day dollars funded a variety of projects/programs (Batheja 2015; Lovell, Barrow, and Wiegand 2008):

- In 2003, state lawmakers took $1.2 billion to help balance the state's budget.
- A decade later, voters approved withdrawing $2 billion to finance a low-interest revolving loan program for water infrastructure projects. (Ramsey and Satija 2013)
- During recent legislative sessions, lawmakers have also suggested using the rainy day fund to pay down state debt, to make the state's retirement systems more solvent, and to improve college affordability. (Garrett 2015)

The state's economy also benefits from the thousands of industry-related and supported jobs throughout Texas. Jobs range from those directly involved in production to secondary and tertiary functions such as pipeline maintenance, wholesalers, field servicing companies, and petrochemical manufacturing (Tables 6.16 and 6.17).

The downturn in prices has harmed the Texas oil and gas industry, although production has remained fairly steady (Gronewold 2015; Jean 2015). Approximately 72,000 jobs were lost in 2015, and industry is

Table 6.16 Oil and Gas Industry Jobs in Texas

Sector	*Direct*	*Indirect*	*Total*
Crude petroleum and natural gas extraction	99,400	215,453	314,853
Natural gas liquid extraction	3,708	8,036	11,743
Drilling oil and gas wells	47,231	84,334	131,566
Support activities, oil/gas operations	134,892	249,324	384,216
Natural gas distribution	8,356	37,476	45,832
Petroleum refineries	20,756	552,670	573,426
Petrochemical manufacturing	15,935	371,531	387,465
Oil and gas field machinery and equipment	54,300	151,453	205,753
Petroleum merchant wholesalers	17,188	19,774	36,962
Pipeline transportation	16,296	116,520	132,816
Total	**418,060**	**1,806,572**	**2,224,632**

Source: Texas House Committee on Business & Industry (2015)

Table 6.17 Employment Benefits

Job Category	*Directly Employed*	*Average Wage*	*Economic Value*
Crude petroleum and natural gas extraction	99,400	179,919	17,883,948,600
Natural gas liquid extraction	3,708	162,368	602,060,544
Drilling oil and gas wells	47,231	105,460	4,980,981,260
Support activities, oil/gas operations	134,892	89,718	12,102,240,456
Natural gas distribution	8,356	113,519	948,564,764
Petroleum refineries	20,756	132,096	2,741,784,576
Petrochemical manufacturing	15,935	121,876	1,942,094,060
Oil and gas field machinery and equipment	54,300	96,762	5,254,176,600
Petroleum products merchant wholesalers	17,188	104,202	1,791,023,976
Pipeline transportation	16,296	143,156	2,332,870,176
Total, annual average	418,060	120,986	50,579,407,160

Source: Texas House Committee on Business & Industry (2015)

projected to shed another 40,000 in 2016 (Hendricks 2016). The economic impacts of low prices are especially acute in the city of Houston, which has lost nearly 20,000 industry jobs during the summer months of 2015 (Sixel 2015). It should be noted that compared with previous bust cycles, Texas's economy is considerably more diversified, which has moderated some of the negative fiscal impacts of the recent downturn, as shown in Figure 6.4.

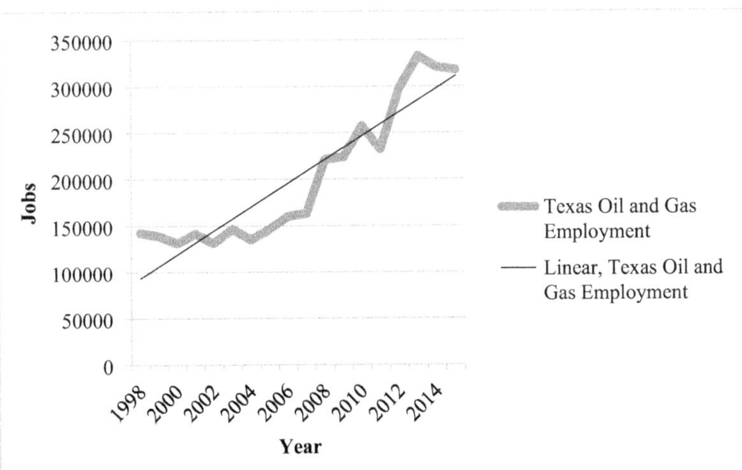

Figure 6.4 Oil and Gas Extraction Jobs
Source: BEA (N.d.)

Local Economic Benefits

The revenues from oil and gas production also support local governments through additional tax revenues and job creation/support. And, oftentimes, it is enough revenue to offset additional costs incurred via production. In FY 2012, for example, Texas's local governments collected approximately $2.3 billion in property taxes from oil and gas firms. Of that, school districts received $1.5 billion; counties, $508 million; cities, $158 million; and other local governments (special districts, etc.), $204 million (Raimi and Newell 2014). The expansive Dallas-Fort Worth region, located atop the Barnett Shale, exemplifies the economic benefits of oil and gas production (Table 6.18). Metroplex leaders in 2008 projected that oil and gas production would add approximately $5 billion to the regional economy and would create between 83,000 and 108,000 permanent jobs by 2030.

Local economic benefits also include increases in taxable sales; an example from the Dallas-Fort Worth area is included in Table 6.19.

Table 6.18 Employment and Payroll in Mining, Quarrying, and Oil and Gas Extraction—Dallas-Fort Worth Metroplex

Year	*Number of Establishments*	*Paid Number of Employees*	*First-Quarter Payroll (thousands)*	*Annual Payroll (thousands)*
2014	1,255	24,289	593,709	2,579,283
2011	1,121	21,642	482,657	2,101,535
2008	1,101	18,812	612,168	2,028,082
2005	902	Between 10,000 to 24,999 employees	Data not available	Data not available

Source: U.S. Census (2015)

Table 6.19 Dallas-Plano-Irving—Mining/Quarrying/Oil and Gas Extraction Sales Data—Pre- and Post- Shale Boom

Dallas-Plano-Irving	*Year*	*Quarter*	*Gross Sales*	*Amount Subject to State Tax*
	2005	1	61,817,359	75,059,622
	2005	2	80,292,742	99,141,699
	2005	3	93,478,348	98,106,063
	2005	4	445,151,254	109,708,496
Total			**$680,739,703**	**$382,015,880**

Dallas-Plano-Irving	*Year*	*Quarter*	*Gross Sales*	*Amount Subject to State Tax*
	2006	1	113,572,267	134,524,436
	2006	2	137,004,263	136,716,574
	2006	3	168,068,352	132,001,388
	2006	4	177,788,141	139,439,117
Total			**$596,433,023**	**$542,681,515**
	2013	1	1,287,576,418	547,132,658
	2013	2	1,949,949,234	504,524,170
	2013	3	2,172,336,294	533,759,521
	2013	4	1,495,084,710	544,177,825
Total			**$6,904,946,656**	**$2,129,594,174**
	2014	1	1,455,229,840	587,414,134
	2014	2	1,963,698,346	673,546,100
	2014	3	2,297,961,783	747,489,846
	2014	4	1,956,766,373	735,600,662
Total			**$7,673,656,342**	**$2,744,050,742**

Source: Texas Comptroller of Public Accounts N.d.

Costs

Similar to the debates in Colorado and Ohio, oil and gas fights in Texas also center on whether production harms the environment, depletes scarce water resources, and impacts infrastructure. An estimate from the advocacy group Environment America is included in Table 6.20.

Table 6.20 Environmental Impacts of Fracking in Texas

Environmental Harm	*2012 Impact (tons)*
Greenhouse gases or equivalent emitted tons since 2005	40,000,000
Acres damaged 2005 to 2012	130,000
Particulate matter	7,800
Nitrogen oxide	100,000
Carbon monoxide	153,000
Volatile organic compounds	14,000
Sulfur dioxide	300

Source: Ridlington and Rumpler (2013)

Air Quality Impacts

Environmentalists have expressed strong concerns that industry has a harmful impact on air quality. Truck traffic, construction, and other aspects of production, they argue, worsen already poor air quality, which is problematic in the Dallas-Fort Worth and San Antonio metropolitan regions (Song, Morris, and Hasemyer 2014). Rawlins (2014) for example, citing an Energy Task Force Committee (organized by the Texas Department of Transportation report), estimated that approximately 1,200 truck deliveries are needed before a single gas well can be brought into its productive phase, as shown in Table 6.21. Leaks and other fugitive emissions from compression units, engines, and oil tanks; well drilling/completing, and fracturing are also possible and can further degrade air quality (Armendariz 2009).

Table 6.21 Estimated Truck Trips and Oil and Gas Production

	Construction—Well Pad	*Trips Made During Operation*	*Total Truck Trips*
One well	1,184 loaded diesel trucks	353 truck trips are required during production per well	1,537
One well pad (six wells)	1,184 loaded diesel trucks	Approximately 2,118 truck trips required per well pad (353 × 6)*	3,302

*Pads can have up to 36 wells.

Source: Rawlins (2014)

Environmentalists also cite the TCEQ's poor commitment to air quality as cause for alarm. They highlight TCEQ's lackluster monitoring and data-gathering abilities (News 4 2015):

- Through early 2014, the TCEQ had installed five permanent air monitor stations, located primarily around the perimeter of the Eagle Ford Shale. The Eagle Ford covers more than 20,000 square miles.
  (Song, Morris, and Hasemyer 2014)

- In 2015, the TCEQ placed its first monitoring station in the center of the Eagle Ford Shale.
  (Song, Morris, and Hasemyer 2014)

- State law enabled many oil and gas facilities to perform self-audits and also self-report their emissions to TCEQ staff.
  (Song, Morris, and Hasemyer 2014)

- Eagle Ford residents, between 2010 and 2013, submitted 284 complaints to the TCEQ. The agency found evidence of 164 violations but only issued fines in two cases. The highest fine amount was $14,000.

(Song, Morris, and Hasemyer 2014)

- State lawmakers reduced the agency's budget by more than $150 million between 2008 and 2014.

(Song, Morris, and Hasemyer 2014)

- Residents in several Barnett Shale communities have reported cut seals and leaking equipment (tubing, valves, etc.)

(Rawlins 2014)

Policymakers in the cities of Dallas and San Antonio are paying close attention to how unconventional oil and gas extraction impacts air quality as well. City leaders in San Antonio caution that the Eagle Ford development is pushing the city up against federal clean air standards. Peter Bella, the Natural Resources Director for the Alamo Council of Governments, expressed concern that "more immediate concern is all the exhaust from the diesel engines in the thousands of trucks, generators, and compressors used to service the well sites" (Bella quoted by Fehling 2012a). Poor air quality is also a concern for lawmakers in the Dallas-Fort Worth region. Lon Burnam (D, Fort Worth) observed, "Those of us in North Texas (have been) in non-attainment for so many years. We absolutely recognize . . . the negative impact on our air quality" (Burnam quoted by Fehling 2012a; Rawlins 2014).

Environmentalists are quick to identify the potentially harmful impacts to air quality, but recent empirical work is more nuanced. Two recent studies focused on Barnett Shale production found a variety of chemical emissions. However, they were not detectable at downwind sites at distances greater than 100 meters. They also noted that concentrations of volatile organic compounds (VOC) related to oil and gas production were lower than "health-based safe exposure levels" (Bunch et al. 2014, Zielinska, Campbell, and Samburova 2014).

Water Availability Impacts

Since 2010, oil and gas producers have, on average, consumed approximately 25 billion gallons of water, which is a tiny percentage (less than 1 percent) of the state's overall consumption (RRC 2014; Wittmeyer 2013). Oil and gas's water 'footprint' also compares favorably with other energy types. University of Texas Bureau of Economic Geology Senior Research Scientist Bridget Scanlon noted, "For every gallon of water used to produce gas, Texas saved 33 gallons of water compared with electrical generation via coal or nuclear fuel" (Scanon as quoted by King 2015).

The impact on nearby communities is highly variable and depends on the geological formation, location, other demands, the presence or absence of drought conditions, the type of well, vertical or horizontal drilling, and the type of chemicals used:

- The average Barnett Shale, Texas, well required 2.8 million gallons.
(Nicot and Scanlon 2012)

- The average Eagle Ford Shale, Texas, well required 4.3 million gallons.
(Nicot and Scanlon 2012)

- The average Haynesville Shale, Texas, well required 5.7 million gallons.
(Nicot and Scanlon 2012)

Despite using a small percentage of the state's overall water, oil and gas production has place-specific impacts (Goldenberg 2014, 2013; Hardwick and Schnyder 2013). Between 2011 and 2013, for example, production limited the water available in 30 West and South Texas communities:

- In Barnhart, Texas, the town well went completely dry.
- In rural McMullen County, located above the Eagle Ford Shale, water consumption by oil and gas operators in 2012 exceeded the county's total withdrawals in 2011.

(Satija and Aaronson 2014)

Activists also cite spills as a cause for alarm. During the spring of 2015, for example, a 42,000-gallon spill forced the evacuation of approximately 100 homes in Arlington, Texas. Spills have also taken place in more rural regions. In 2015, a wastewater spill in the Hagerman National Wildlife Refuge caused $154,000 in property damage and killed 84 mature oak, cottonwood, and hackberry trees (Soraghan 2015a). Jackson et al. (2014) noted that there are confirmed examples of groundwater contamination in Texas, often the result of poor well construction. The researchers explained that contamination at the Texas site was due to the company cementing "very near the surface and deep underground, but they put no cement for 4,000 feet in between." In effect, the cement gap enabled gases to migrate up and down similar to a 'chimney' (Jackson et al. 2014).

Concerns About Poor Enforcement

Environmentalists contend that the RRC is captured by industry and too eager to promote development even at the expense of environmental protection and public health, as reflected in Table 6.22 (Begos 2014; Davis 2012). Activists highlight that in FY 2009, the agency identified

Table 6.22 RRC Enforcement

Fiscal Year	*Oil and Gas Inspections*	*Alleged Violations Identified*	*Complaints Received*	*Fines*
2016	124,299	41,867	616	$3,697,434
2015	134,484	61,189	680	$3,603,050
2014	130,812	62,385	754	$2,616,723
2013	125,878	55,030	791	$1,178,631
2012	118,484	55,960	727	$1,965,019

Source: RRC (2016a)

approximately 80,000 violations. It, however, only pursued enforcement actions in small percent of those violations (Soraghan 2012, 2011; RRC 2015, 2014a, 2014b).

In addition to its enforcement records, critics point to the RRC's governance structures as problematic. In its 2011 Final Report, the Texas Sunset Advisory Commission characterized RRC's enforcement process as lacking a clear method for pursuing enforcement action in a way that is "based on a consistent measure of severity or pattern of repeat offenses" (Sunset Advisory Commission 2011, 2, 3). Critics also highlight the agency's system of electing commissioners. State Senator Rodney Ellis, a Democratic member of the Natural Resources Committee, observed that the existing method for electing RRC commissioners may give rise to the perception of potential conflicts of interest, "sitting there for a six-year term, being able to raise unlimited amounts of money from the industry that they regulate . . . clearly is a perception problem" (Ellis as quoted by Barer 2013).

Public Safety and Boom/Bust

The boom and bust nature of oil and gas economics has led to additional challenges for local governments (Block 2014). The experiences of the Texas city of Three Rivers are telling. At the height of the boom, the town's revenues and property values swelled. City leaders credited the surging oil and gas industry for paying for the city's new high school and athletic facilities, and for supporting four new hotels. Three Rivers Mayor Sam Garcia also identified some of the costs of development: "The city's small (ten-person) police department is encountering a surge of traffic calls, break-ins and are even dealing with increasing prostitution" (Garcia quoted by Jervis 2014).

The city of Karnes City, located atop the Eagle Ford Shale, has also experienced the costs and benefits of the energy boom. During the height of operations, Karnes City saw an influx of new workers, sales tax revenues, increasing vehicle and home sales, and growth in its retail sector. However, as oil and gas prices have declined, the city's economy has slowed. In the summer of 2015, Vi Malone, the Karnes County treasurer, described the region's economic outlook as "everybody is waiting for doomsday" (Malone as quoted by Krauss 2015).

Summary on Texas's Intergovernmental Issues

Texas law divides authority between Texas cities and state regulatory agencies. With the enactment of HB 40, the state has limited local authority to regulate oil and gas production. The state has also implemented a "commercially reasonable" standard as its baseline for evaluating the legality of a local action. Even with HB 40, Texas's communities have been and will likely remain active policymaking venues when it comes to reconciling quality of life, economic development, and environmental protection issues. Cities have enacted ordinances that touch upon a wide variety of oil and natural gas surface impacts such as limiting site access and the hours in which drilling/venting/flaring may take place, prohibiting specific locations, and requiring security features and aesthetics.

Notes

1 Three gubernatorial appointees set the agency's general priorities, and an executive director oversees the agency's day-to-day operations (TCEQ 2014, 2014a).
2 Three elected commissioners lead the RRC (RRC 2014; University of Texas Libraries 2014).
3 Once the applicable fees are paid, firms supply regulators with lease, well depth, and location (distance to nearest city, well, etc.) information (Rahm 2011; RRC 2014; Texas State Library Archives 2009). The RRC also requires that operators disclose well bore data, including acres leased, directional or vertical drilling, leasing rights (i.e., single or multiple owners), casing, cementing, well integrity, completion, and disposal plans (Davis 2012). The RRC reserves the right to shutter a well if regulators determine the operator is violating RRC policies (Allen 2014; RRC 2014; Rutherford 2014).

7 Local Oil and Gas Stories

Intergovernmental dynamics are influenced by the experiences, whims, relationships, and idiosyncrasies of office holders. They are further complicated by context-specific factors such as the presence of a focusing event (Davis 2014; McGuire 2006; Stever 2005; Wright 1978). For some cities, these external events and other unique characteristics do not lend themselves to statistical analyses, but are critical to understanding intergovernmental conflict (Anderson 1960; Klyza and Sousa 2007; Wright 1978). Given that cities experience the costs and benefits of oil and gas production differently, an important step to understand defiance is through examining the experiences of local governments. The 'local experience' includes both cities defying (or that have defied) the state and those that are actively addressing oil and gas development but within state rules. Even those cities whose policies fit within applicable state frameworks may identify concerns that are crucial to avoiding intergovernmental conflicts in the future. In short, by gaining a deeper understanding of the sources of defiance and/or local anxiety, it then becomes possible to identify more effective governing arrangements.

Communities weigh and evaluate the politics of local defiance and that of oil and gas development differently. Despite differing values and state frameworks, a number of common themes related to the challenges of natural resource governance have emerged, some of which are captured in the passages in the following:

- Roy Otto, city manager of Greeley, Colorado, noted, "We'd (the city) like to be a part of figuring out how the resources can be developed to the fullest potential for the benefit of the world and ourselves."

  (Otto as quoted by Finley 2013)

- Fort Collins, Colorado, Mayor Pro Tem Kelly Ohlson criticized the state's Democratic Governor John Hickenlooper's support for fracking as "I believe the Governor should spend his time protecting the health and safety and welfare of citizens of Colorado rather than acting like the chief lobbyist for the oil and gas industry."

(Ohlson as quoted by Stokols 2013)

- Munroe Heights, Ohio, Mayor Frank Larson described his city's attempt to restrict oil and gas development as, "It's not about oil and gas. It'd be simple if it was. . . . It's not about fracking and it's not about gas wells. It's about home rule."
(Larson as quoted by Higgs 2014)

- Speaking about the nature of boom and bust cycles, James O'Reilly, a council member for the city of Wyoming, Ohio, said, "When the gas drillers see Ohio in their rear view mirror, we'll want to know that we did everything we could to protect our communities from their harms."
(Reilly as quoted by Ouzts 2015)

- Dallas, Texas, Mayor Mike Rawlings cautioned, "We need to think carefully when we vote on actions to not put ourselves in the courthouse." The statement followed the city's passage of a 1,500-foot setback for natural gas wells (Rawlings as quoted by Wilonsky 2014). He added that he believes the city will prevail in its dispute with Trinity East (Malewitz 2013a). However, it is highly likely that the city's setback distance would fail to meet the commercially reasonable standard enshrined in state law.
- Denton, Texas, Councilman Kevin Roden, in reference to House Bill 40, which banned municipal bans and imposed other restriction on local regulatory authority, noted that for some lawmakers there is a belief "that big state government knows better than the citizens of a city. I just think—conservative or liberal—that is something you don't do in Texas."
(Roden as quoted by Gold 2015)

- Texas Governor Greg Abbott (R) warned that HB 40 was needed to protect industry from "the heavy hand of local regulation."
(Abbott as quoted by Goodwyn 2015)

Sampling Process

The communities in this chapter were selected based on a purposive quota sampling technique. As part of this strategy, cases must meet specific criteria and are not randomly selected (Tashakkori and Teddlie 2003; Patton 2002). As applied to local oil and gas policies, a multi-step approach was adopted, as follows:

1. Cities must be located in states with ongoing oil and gas development.
2. Cities must be located in counties with ongoing oil and gas development.

Four cities that met the foregoing criteria per state (Colorado, Ohio, and Texas) were selected (Table 7.1). Although data are non-random, which

Table 7.1 Municipal Case Studies, by State

Colorado	*Ohio*	*Texas*
Longmont	Munroe Falls	Arlington
Fort Collins	Youngstown	Denton
Greeley	Broadview Heights	Fort Worth
Loveland	Akron	Mansfield

Source: Author generated

limits generalizability, the case studies do illuminate many of the costs and benefits that local governments must confront with oil and gas operations. Importantly, they offer deeper insights into the potential motivations that underpin local challenges to the state's preemptive authority.

City of Longmont, Colorado

Longmont is a council-manager city located north of Denver; it has a population approaching 90,000. The city's governmental structure consists of seven elected and non-partisan council members: three members who represent specific geographical districts; three who serve as at-large members, and a mayor. An appointed city manager directs the city's administrative operations (City of Longmont 2013). The city's per capita income and educational attainment levels are near the state average (Table 7.2). The city of Longmont has committed itself to reducing and preventing pollution, developing and relying on green-energy resources, and achieving gains in energy efficiency. The city also offers energy efficiency loans and rebates to property owners and single-stream recycling for residents

Table 7.2 Socio-Economic Characteristics of Longmont

Characteristic	*Longmont*	*Colorado*
Homeownership rate, 2009–2013	62.1%	65.4%
Housing units in multi-unit structures, 2009–2013	28.7%	25.9%
Median value of owner-occupied housing units, 2009–2013	$238,900	$236,200
Per capita income in past 12 months (2013 dollars), 2009–2013	$29,209	$31,109
Median household income, 2009–2013 INC110213	$58,698	$58,433
Persons below poverty level, 2009–2013	14.7%	13.2%
Education, bachelor's degree or higher	37.2%	37%

Source: U.S. Census Bureau (2016g)

(City of Longmont 2013a). In 2016, the city's well inventory consisted of 16 abandoned and/or plugged wells, 19 abandoned locations (well locations that were plotted but never drilled), and three dry abandoned wells (Denver Post Editorial 2012; City of Longmont 2016, 2013b).

Formal challenges to the state began in 2012, the origins of defiance, however, trace back to the 1980s (Antonacci 2016; Jaffe 2013, 2013a; Rochat 2012, 2012a, 2012b):

- In 1982, Calvin Petroleum Corporation purchased the Rider Well, which was located in unincorporated Boulder County and not located near any school (at the time).
- In 1986, TOP Operating purchased the Rider Well. At the time it was common practice to use open pits to store oil and gas's wastewaters. These pits were then subsequently covered.
- Operators fracked at the Rider Well site (did not horizontally drill) to extract oil and gas in 1983 and again in the late 1990s.
- In the 1990s through early 2000s, the city grew from 43,500 residents to more than 76,000. There was very little drilling in Longmont at the time (most development was located east of the city in rural portions of Weld County).
- In 2006, Engle Homes, a property developer, notified the COGCC that groundwater tests showed benzene levels of 491 parts per million. The state had established a limit of five parts per million.
- In 2008, COGCC regulators fined TOP Operating for failing to remediate quickly enough.
- In 2011, Longmont residents began to question if the Rider Well should be located so close to occupied structures (350 feet was the COGCC setback distance).

The city's fracking-related defiance peaked in July 2012, when city lawmakers banned drilling in residential areas (passed on a 4–3 vote). In addition to the ban, staff members negotiated (in a 6–1 vote) a comprehensive contract with TOP Operating. The contract stipulated a number of agreements for when, where, and how TOP would drill:

- TOP would drill on 11 sites on city-owned property.
- TOP would conduct mandatory water sampling, testing, and monitoring of areas near drill sites.
- The contract also stipulated that TOP Operating abide by a 750-foot setback between wells and any occupied building.
- The city would purchase the Rider Well so that TOP could shut it down.
- The city would pay for a replacement well.

However, before the agreement could be finalized, Longmont voters in November 2012 banned fracking by a 60–40 margin (Antonacci 2016;

Jaffe 2013, 2013a; Rochat 2012, 2012a, 2012b). It should be noted that the city's ban did not include other extraction methodologies (Denver Post Editorial 2012; City of Longmont 2016, 2013b). TOP and the city agreed to suspend the agreement until the legal questions surrounding the fracking ban were resolved.

The Colorado Oil and Gas Association sued the city in December 2012, seeking to block many of the city's policies towards oil and gas production. The COGCC later joined this COGA lawsuit. The COGA suit was in addition to a lawsuit already filed by the COGCC. The COGCC (which COGA joined) suit charged that Longmont's oil and gas ordinance and its ban on drilling in residential zones, monitoring policies, and setback distances were preempted by existing state rules. In reference to the COGA-filed lawsuit, Matt Lepore, the executive director of the COGCC, noted, "The COGCC did not initiate this lawsuit or this process . . . that said, the COGCC does believe Longmont's ban on hydraulic fracturing is contrary to state law, and we believe clarity from the courts on this matter is important" (Jaffe 2013). The state dropped its lawsuit as part of the governor's state-local task force compromise. In July 2014, Boulder District Court Judge Dolores Mallard nullified Longmont's fracking ban (Antonacci 2015; Denver Post Editorial Board 2014; Rochat 2014).

The decision was quickly appealed to the State Supreme Court. The city articulated two principles as its legal argument. First, city leaders argued that the ability to limit the locations of wells is protected by its home rule and land use authority (Rochat 2013). Longmont Mayor Dennis Coombs explained that local governments "already have the right to restrict heavy industrial uses from residential zones, and that oil and gas drilling shouldn't be any different" (Jaffe 2012). He also believed that the city's policies do not substantially impede or conflict with COGCC regulations (Jaffe 2013, 2012). Additionally, because state rules do "not affirmatively or expressly authorize fracking," a city policy that bans it does not actually prohibit an activity that state law permits. As such, there is no operational conflict between the COGCC's regulations and the city's rules (Rochat 2013).

In 2016, the State Supreme Court ruled against the city and found that state law superseded Longmont's ban. With the legal issues clarified, city staff and TOP Operating have again engaged in negotiations with how to regulate production within the city. In late July 2016, the city and TOP announced that they had agreed to close the Rider Well. Mayor Coombs said, "This is another example of the great working relationship we have with the oil and gas operators in our community." TOP officials have also expressed a commitment to work with the city and to shut down the Rider site before the school year. They have also agreed to engage in reclamation work around the well, which is not required by existing COGCC rules (City of Longmont 2016a).

City of Fort Collins, Colorado

The city of Fort Collins is located at the northern edge of Colorado's highly populated Front Range. City government consists of six non-partisan council members, each elected by district, and a mayor (City of Fort Collins 2016). Each council member serves a four-year term; the mayor serves for two years. A city manager oversees the city's administrative side. The city is also home to Colorado State University. Since 2000, city staff has conducted a greenhouse gas inventory, enacted energy efficiency policies and incentives, implemented a voluntary environmental program, authored greenhouse gas mitigation plans, and attempted to reduce vehicle miles traveled within the city (Mosier and Fisk 2013). In 2013, city officials tallied seven producing wells, mostly clustered in the northeastern portions of the city (City of Fort Collins 2013; Finley 2013). The socio-economic characteristics of the city are provided in Table 7.3.

Table 7.3 Socio-Economic Characteristics of Fort Collins

Characteristic	*Fort Collins*	*Colorado*
Homeownership rate, 2009–2013	55.5%	65.4%
Housing units in multi-unit structures, 2009–2013	32.1%	25.9%
Median value of owner-occupied housing units, 2009–2013	$247,800	$236,200
Per capita income in past 12 months (2013 dollars), 2009–2013	$28,729	$31,109
Median household income, 2009–2013	$53,780	$58,433
Persons below poverty level, 2009–2013	18.6%	13.2%
Education, bachelor's degree or higher	51.9%	37%

Source: U.S. Census Bureau (2016d)

Local defiance in Fort Collins began with more persuasive measures aimed at providing residents and lawmakers with additional information gained through research on oil and gas's impacts on residents' health and property values. Staff concluded that they lacked the requisite data to draw any firm conclusions or to make any policy recommendations. The studies, however, did find that because of the limited nature of oil and gas operations within the city, industry posed minimal or no health risk to city residents. The city of Fort Collins also reported that between 2011 and January 2016 operators recorded nine spills/leaks within the city's Growth Management Area. The majority of spills were small, and operators recovered most of the escaped fluids almost immediately (City of Fort Collins 2016a). A list of spills is available at www.fcgov.com/oilandgas/

In March 2012, the city council temporarily banned fracking and blocked the storage of fracking fluids within city limits. Council members explained that the moratorium would enable them the time they needed to study the issue, develop local regulations, allow for state legislative and/or regulatory changes, study fracking's public health and environmental impacts, and give the city enough time to contribute to the COGCC's rulemaking processes. The moratorium also offered the city time to work with Prospect Energy to develop site-specific plans. Later in 2012, the city announced that it had agreed with Prospect Energy and as part of the agreement the city had exempted the company from its moratoria and ban (City of Fort Collins N.d.; Weinheimer 2013).

Opposition to fracking reached its zenith during the fall of 2013 when activists succeeded in placing the question of a fracking moratorium on the ballot. Nearly 58 percent of Fort Collins's voters supported the moratorium's extension for five years and banning fracking fluid storage (Hindi 2013). Before the vote, city council members issued a resolution opposing the moratorium's extension, citing concerns over its legality.

Following the moratorium's passage, city officials began implementing the new law and preparing for the anticipated legal fallout (Hindi 2013). In December 2013, the Colorado Oil and Gas Association [COGA] sued the city, charging that it violated existing COGCC rules. In 2014, Larimer County District Court Judge Gregory M. Lammons struck down the city's moratorium, explaining that it operationally conflicted with COGCC rules (*Colorado Oil and Gas Association, et. al v. city of Fort Collins, et. al* (13CV31385) (Handy 2014). In 2016, the State Supreme Court held that COGCC rules preempted the city of Fort Collins's moratorium. The decision, however, clarified that the use of moratoria are appropriate land use tools if they are "interim measures that are, by their very nature, of limited duration and are designed to maintain the status quo pending study and governmental decision making."

City of Greeley, Colorado

The city of Greeley operates as a home rule, council-manager city and is also located in Northern Colorado. An elected mayor, four districted council members, and two at-large members establish policy and set the city's goals. An appointed city manager leads the city's daily operations. It is home to the University of Northern Colorado, and approximately 26 percent of its residents hold a bachelor's degree or higher. The city's median household income ($46,272) is significantly less than the state average, as shown in Table 7.4 (City of Greeley 2016).

The city of Greeley has a long history of natural gas and oil drilling. It is home to 427 wells, with another 1200 projected in its urban growth area. The city expects its inventory of current and future wells to generate millions of dollars in revenues (Healy 2013). In 2012, for example,

Table 7.4 Socio-Economic Characteristics of Greeley

Characteristic	*Greeley*	*Colorado*
Homeownership rate, 2009–2013	56.6%	65.4%
Housing units in multi-unit structures, 2009–2013	31.2%	25.9%
Median value of owner-occupied housing units, 2009–2013	$166,300	$236,200
Persons per household, 2009–2013	2.68	2.53
Per capita income in past 12 months (2013 dollars), 2009–2013	$21,845	$31,109
Median household income, 2009–2013	$46,272	$58,433
Persons below poverty level, 2009–2013	22.9%	13.2%
Education, bachelor's degree or higher	25.8%	37%

Source: U.S. Census Bureau (2016f)

city officials estimated that oil and gas operations generated $3.3 million in local revenues and over the next 25 years they believe that industry's financial contributions may surpass $420 million. The city also leases its surplus drinking water to industrial users (City of Greeley 2013). By water leasing, the city leaders added between $4 million and $5 million to its FY 2012 budget (Healy 2012).

The city regulations, operating within the COGCC legal parameters, address a number of drilling's impacts:

- Through its permitting process, the city may impose screening and camouflage standards, require operators to mitigate weeds, and abate other nuisances located on the well pad.
- The city's permits may limit the routes available to the large trucks needed by oil and gas operators, or they may include an impact fee designed to reimburse the city for the costs associated with development.
- The city's permit may also mandate that operators install wildlife passages.
- For each pad or well, planning staff reviews the application and determines which conditions need to be included in the permit.

(City of Greeley 2011)

City of Loveland, Colorado

Compared with the city of Greeley, Loveland officials have been more critical of oil and gas development. But, unlike Longmont or Fort Collins,

Table 7.5 Socio-Economic Characteristics of Loveland

Characteristic	*Loveland*	*Colorado*
Homeownership rate, 2009–2013	64.2%	65.4%
Housing units in multi-unit structures, 2009–2013	23.6%	25.9%
Median value of owner-occupied housing units, 2009–2013	$210,400	$236,200
Households, 2009–2013 HSD410213	28,338	1,977,591
Persons per household, 2009–2013	2.41	2.53
Per capita income in past 12 months (2013 dollars), 2009–2013	$27,878	$31,109
Median household income, 2009–2013	$54,977	$58,433
Education, bachelor's degree or higher	32%	37%

Source: U.S. Census Bureau (2016h)

the city of Loveland decided on a policy course that was much less likely to challenge the state's preemptive authority. The city's charter vests legislative power in nine elected officials. It is divided into four wards, with two members elected per ward, and an elected mayor representing the entire city. An appointed administrator manages the city's day-to-day affairs (City of Loveland 2013). Its 70,000 residents approach the state average in terms of education (32 percent with a bachelor's degree) and income level ($54,977) (U.S. Census 2013c) (Table 7.5). City leaders have employed a triple bottom-line approach (economic, social, and environmental implication) to inform decision-making. The plan, however, does not specifically address climate change or oil and gas policies (City of Loveland 2013a).

As it became more apparent that Anadarko was going to drill within the city, officials decided to collaborate with the company and state regulators (Maher 2013). Early cooperation, according to Anadarko officials, protected the city from future legal actions, protected the interests of mineral owners, and ensured public health. Through its collaboration, the city removed language that banned development in floodplains and avoided placing regulations on produced fluid storage pits and other fracking-fluid disposal practices (Maher 2013).

In March 2013, the city council narrowly passed a comprehensive set of oil and natural gas regulations. The ordinance established a framework that allowed the city to inspect sites for violations, although it acknowledged that the city cannot enforce any conditions that conflicted with applicable COGCC provisions. The law also presented operators with two options should they seek to drill within the city. Operators may meet the standards established by the COGCC but be subject to a local review process that included public hearings before the city's planning commission, impacted neighborhoods, and the city's development team. This option, according to Loveland officials, may take months and include appeals. Under the second option, operators may elect to sign a voluntary agreement(s) with the city in exchange for an expedited local permitting process (City of Loveland 2015; Maher 2013).

Before the enactment of the city's comprehensive ordinance, city officials flirted with a long-term moratorium (Maher 2014). However, it ultimately opted for a set of local standards that were more likely to survive judicial scrutiny (Maher 2013). City councilman Daryle Klassen explained that "other cities will look to us for a template . . . we have nothing but to gain for enacting these for our community," said (Maher 2013). City councilman Hugh McKean reflected that "you cannot have 500 sets of rules for every jurisdiction, every county and every municipality when it comes to an industry that has to operate across Colorado using the same technology" (McKean quoted in Observer Staff 2013).

Other Loveland lawmakers were more critical of the city's approach. Councilman Ralph Trenary stated, "I can't agree to the city of Loveland deciding our policy and programs in that kind of convoluted, manipulated process" (Trenary as quoted by Maher 2013). Councilman Phil Farley equated the participation of the COGCC and industry to the fox guarding the henhouse (Farley as quoted by Maher 2013). Mayor Cecil A. Gutierrez was particularly pointed in his criticism of the state and industry. He suggested that the ordinance was highly influenced by the COGCC and industry and that as a city "we bent over backward to abdicate, acquiesce to the state" (Gutierrez as quoted by Maher 2013). He also identified several local preferences that were stripped from the state-approved standards:

- Language that mandated that operators reduce the vibrations created when drilling within the city
- Language that required operators to disclose the contact information of individuals who would be responsible should there be an on-site emergency
- Language that would have granted the city access to maps showing the locations of well-heads, pipelines, and other equipment used by operators in the case of an emergency
- Language that required operators to grant local authorities access to Material Safety Data Sheets (MSDS) for chemicals used as part of the fracking process; the ordinance does have language that mandates disclosure of fracking chemicals and spill reporting to the Loveland Fire and Rescue Authority
- Language that would have obligated operators to participate or plan citizen education programs about potential hazards related to extraction
- Language that would have strengthened air and water monitoring and testing

Opposition to fracking continued to percolate throughout the community and led to a special ballot initiative. In 2014, residents voted down a proposed two-year moratorium by less than 900 votes (out of 20,000 cast) (Maher 2014).

Following is a summary of sources of defiance in Colorado (Table 7.6):

Table 7.6 Sources of Local Concern in Colorado

Factor	*Longmont (defiance)*	*Fort Collins (defiance)*	*Greeley (alignment)*	*Loveland (alignment)*
Negative impacts to environment	Yes	Yes	No	Yes
Lack of information sharing (emergency response)	NA	No evidence	No evidence	Yes
Focusing events	Yes	Marginal	No evidence	No evidence
Negative impacts to homeowners	No evidence	No	No	No evidence
Concern over state oversight or policies	No evidence	No evidence	No evidence	Yes

Source: Author generated

City of Munroe Falls, Ohio

The city of Munroe Falls is located in northeast Ohio and is led by an elected mayor and seven council members. Four of the members represent the city at large, with the remaining three elected via ward. The mayor oversees the city's daily operations. The city, with a population of 5,500, markets itself as a bedroom community with upscale housing, a high quality of life, numerous parks, and a multitude of festivals and cultural events (City of Munroe Falls 2015). Census data demonstrates that the city's housing inventory, compared with the rest of Ohio, is more likely to be occupied by homeowners (rather than renters) and have a greater median value, as noted in Table 7.7. Residents are also more educated and affluent compared with other areas of the state.

The city's experiences with oil and gas production involve symbolic and legal acts of defiance. Between 2011 and 2012, council members released several resolutions highlighting their concerns related to drinking water contamination, industry's poor environmental record, procedural fairness during ODNR rulemaking, the state permitting-approval process, and the city's general lack of confidence in how the state polices the oil and gas industry. Many of these complaints are noted (City of Munroe Falls N.d.) as follows:

Table 7.7 Socio-Economic Characteristics of Munroe Falls

Characteristic	*Munroe Falls*	*Ohio*
Homeownership rate, 2009–2013	81.9%	67.5%
Housing units in multi-unit structures, 2009–2013	18.7%	22.9%
Median value of owner-occupied housing units, 2009–2013 HSG495213	$162,800	$129,600
Per capita income in past 12 months (2013 dollars), 2009–2013 INC910213	$35,021	$26,046
Median household income, 2009–2013	$64,654	$48,308
Persons below poverty level, 2009–2013	2.0%	15.8%
Education, bachelor's degree or higher	39.0%	25.6%

Source: U.S. Census Bureau (2016j)

- Highlighted that ODNR staff issued approximately 1,400 oil- and gas-drilling permits since the oil and gas renaissance began in 2008, with none being denied on environmental grounds.
- Criticized state leaders for passing House Bill 278, which preempted local control relative to oil and gas production without considering how it may impact existing local actions, ordinances, and preferences. The resolution also stated that Substitute Senate Bill 165 enabled oil and gas operators to effectively ignore local actors and preferences. According to the city, oil and gas firms "have become emboldened by the cover [provided by SB 165] . . . and have begun to take the position that the drillers do not have to comply with any local ordinances."

  (City of Munroe Falls 2012)

- Faulted state lawmakers for failing to require the ODNR to solicit and collect meaningful information from local communities as well as from the Ohio Environmental Protection Agency about the risks posed by operations in environmentally sensitive areas. In fact, the city noted, "The ODNR was not required to do any sort of specific groundwater risk or environmental assessment."

  (City of Munroe Falls 2012)

- Voiced concerns over the ODNR's record in protecting Ohio's drinking water. The resolution noted that between the mid-1970s and 2012, ODNR officials investigated approximately 900 drinking water contamination complaints linked to oil and gas development. Recorded that ODNR staff admitted that the state has been lucky that many of these complaints originated with private water supplies, rather than public water supplies.

  (City of Munroe Falls 2012)

- Cited the experiences of Pennsylvania, West Virginia, and New York that fracking wastewater issues are unresolved and pose a threat to drinking water resources.

(City of Munroe Falls 2012)

The city's legal challenges erupted when a resident leased property to Beck Energy for a vertical well. The state issued a permit in February 2011, which stipulated 67 conditions (38 tied to the site's classification as a municipal wellhead protection area and 29 others that addressed the site as an urbanized area). Beck began site preparations without a local permit. In response, the city issued a stop-work order and filed a lawsuit seeking to block Beck from drilling the well (*State ex rel. Morrison v. Beck Energy Corp.*, 143 Ohio St.3d 271, 2015-Ohio-485).

The city's permitting process imposed a number of conditions on operators before drilling. First, as part of its general zoning powers, the city required Beck Energy to receive a zoning certificate issued by the zoning inspector before constructing or excavating. The certificate requires approval from the planning commission, city council, and the zoning inspector. Second, under city law, Munroe Falls prohibited Beck from drilling (for oil, gas, or other hydrocarbons) until it could demonstrate that it met all conditions outlined in Chapter 1329 of the ordinance (City of Munroe Falls 2016):

- Munroe Falls Codified Ordinances 1329.04 and 1329.06 required applicants to pay a fee of $800 and a $2,000 performance bond—both due at the time of filing.
- Munroe Falls Codified Ordinance 1329.05 required a public hearing to be conducted at least three weeks before drilling.
- The applicant must also schedule the hearing and inform all property owners within 1,000 feet of the wellhead site of the meeting and proposed well.
- City code also stipulated that parties found in violation of Chapter 1329 were guilty of a first-degree misdemeanor and subject to fines and possible jail time.

As outlined in Chapter 5, the State Supreme Court found that state laws preempt the city's oil and gas ordinance (Cama 2015b). Following the decision, the city council members repealed the oil and gas ordinance (Smith 2015a).

City of Broadview Heights, Ohio

Broadview Heights's policies have also challenged the ODNR and state leaders. Located south of Cleveland, the city is home to 20,000 residents. Its legislative power is vested in a seven-member council (three members serve as at-large representatives and the other four represent specific

Table 7.8 Socio-Economic Characteristics of Broadview Heights

Characteristic	*Broadview Heights*	*Ohio*
Homeownership rate, 2009–2013	81.8%	67.5%
Housing units in multi-unit structures, 2009–2013	18.8%	22.9%
Median value of owner-occupied housing units, 2009–2013	$213,700	$130,800
Households, 2009–2013	7,681	4,557,655
Per capita income in past 12 months (2013 dollars), 2009–2013	$37,115	$26,046
Median household income, 2009–2013	$75,882	$48,308
Persons below poverty level, 2009–2013	6.6%	15.8%
Education, bachelor's degree or higher	47.7%	25.6%

Source: U.S. Census Bureau (2016b)

geographical districts) (City of Broadway Heights 2015). An elected mayor manages the city and serves as the chief law enforcement officer (City of Broadway Heights 2015a). The city is also relatively wealthy and well educated, as shown in Table 7.8.

Between 2000 and 2011, operators drilled more than 80 wells (mostly vertical) inside the city's territorial limits. Initially, the city welcomed the development. In 2011, for example, it leased space for 15 vertical wells, which generated $163,000 in additional revenues for the city. The city's relationship with industry and the state, however, began to sour when the ODNR issued a mandatory pooling order in 2010 (Karpus 2010). ODNR issued the order after GonzOil sought to drill a well within the city, but refused to comply with an ordinance that required a buffer of at least 500 feet between the well and any non-participating property. City law also required the operator to pay the city an up-front fee of $10,000. After determining that it would not comply with the city ordinance, GonzOil received the pooling order from the ODNR (Karpus 2010). If the state issues a mandatory pooling order, an operator may drill on the property owned by non-cooperating property owners.

Tensions over the pooling order and development continued. Throughout 2011 and 2012, activists and local officials raised concerns over fluid spills (a neighboring community experienced a 200-gallon spill in 2012), local control, and air pollution. Robert Stefanik, mayor of North Royalton (a neighboring community), described that after drilling began, residents started noticing spills and accidents. During one accident, fluids leaked into a nearby creek, which flowed into the city of Broadview Heights (Sandrick 2013).

City officials have also weighed in on discussion over local oil and gas authority. The mayor of Broadview Heights, Sam Alai, described that it is "absurd" that in Ohio, cities can require building permits for new home construction and improvement projects, but have no authority to influence the placement of wells (Alai as quoted by Spector 2012). He added that he is concerned about the placement of wells so close to homes and in residential areas (Spector 2012).

In 2012, activists placed a charter amendment (Issue 29) on the city's ballot, which would have banned all future wells and wastewater sites from Broadview Heights. The amendment passed in November 2012 with nearly two thirds of voters supporting it (6,211 for versus 3,188 against) (Rozboril 2012). The amendment challenged the state in several ways (City of Broadview Heights 2012):

1. The amendment declared that Broadview Heights residents hold "an inalienable and indefeasible right to self-governance in the community where they reside."
2. The amendment prohibited oil and gas extraction within city limits but allowed existing wells to continue to produce hydrocarbons (although they may not change extraction processes).
3. The amendment called for the city to refuse to recognize permits, licenses, privileges, or charters issued by higher levels of governments, including those from the ODNR that conflicted with the language of the charter ordinance.

In 2015, Cuyahoga County Common Pleas Judge Michael K. Astrab voided the city's Community Rights Amendment. Judge Astrab concluded that state lawmakers authorized the ODNR to be the "sole and exclusive authority to permit, locate, space, and regulate oil and gas wells." As such, ODNR regulations preempted city policies that restricted the location of future wells (an activity permitted by the state). The judge also found that state laws prevented cities and townships from enacting ordinances that discriminate, impede, or obstruct oil and gas development. Finally, the ruling characterized the city's ordinance as violating the property rights of Bass Energy Co. Inc. of Fairlawn, Ohio, and Ohio Valley Energy of Austintown, Ohio (Sandrick 2015). As of 2016, the city had not yet announced if it was going to appeal the decision.

City of Youngstown, Ohio

Youngstown, Ohio, is an older industrial community of 65,000, led by an elected mayor and seven districted council members. City leaders are firmly committed to economic development, and many have spoken in favor of development in the Utica formation. They have also dedicated resources to reducing litter, abating abandoned properties, and recycling (City of Youngstown 2016, N.d.; Zremski 2014). The city's socioeconomic characteristics are presented in Table 7.9.

Table 7.9 Socio-Economic Characteristics of Youngstown

Characteristic	*Youngstown*	*Ohio*
Homeownership rate, 2009–2013	57.7%	67.5%
Housing units in multi-unit structures, 2009–2013	20.7%	22.9%
Median value of owner-occupied housing units, 2009–2013	$45,400	$129,600
Per capita income in past 12 months (2013 dollars), 2009–2013	14,742	$26,046
Median household income, 2009–2013	$24,361	$48,308
Persons below poverty level, 2009–2013	37.4%	15.8%
Education, bachelor's degree or higher	11.5%	25.6%

Source: U.S. Census Bureau (2016k)

Youngstown's experiences with unconventional oil and gas development have involved clashes among grassroots anti-fracking activists, city leaders, clergy, and industry officials (Hartman 2014). Between 2009 and 2015, city residents voted on several proposed fracking bans. Each proposal was defeated (Skolnick 2015):

- In 2014, supporters defeated the proposed ban by 8.3 percentage points.
- In 2013, supporters defeated the proposed ban by 9.7 percentage points.
- In 2011, supporters defeated the proposed ban. This result was notable because it followed a series of high-profile earthquakes in the region.

(Niquette 2012a; Skolnick 2015)

The 2014 ballot effort exemplified the complicated politics associated with managing oil and gas production. Fighting against the ban was a bipartisan coalition of businesses and organized labor interests. Local religious leaders also lobbied against the ban. Reverend Gregory Maturi of St. Dominic's Catholic Church, an inner-city Youngstown faith community, highlighted many of the economic benefits associated with oil and gas development: "This is really to help the poor and needy. It's about helping people get jobs" (Maturi as quoted by Zremski 2014).

In Youngstown, nearby oil and gas production has supported new jobs and industry investment. Vallourec, a French company specializing in piping, invested $1.1 billion in a plant intended to supply oil and gas producers with piping (Zremski 2014). Development's costs, by comparison, were more scattered across the region. In 2013, there were only six producing wells in Mahoning County and none in Youngstown; see Table 7.10.

Table 7.10 City of Youngstown Area Wells, 2014

Operators	*Township*
Chesapeake Exploration, LLC	Milton
CNX Gas Company, LLC	Ellsworth
CNX Gas Company, LLC	Jackson
CNX Gas Company, LLC	Jackson
Hilcorp Energy Company	Poland
Halcon Operating Company, Inc.	Jackson
Hilcorp Energy Company	Poland
Hilcorp Energy Company	Poland
Hilcorp Energy Company	Poland
Hilcorp Energy Company	Poland
Hilcorp Energy Company	Poland
Hilcorp Energy Company	Poland
Halcon Operating Company, Inc.	Jackson

Source: ODNR (2016a)

In 2013, the city announced that it would seek bids for its mineral rights underneath several city-owned properties (non-park properties) for between $5,000 and $7,500 an acre, plus signing bonuses (Geewax 2013). Youngstown Councilman John Swierz explained his support for the bid as "fracking provides a solution" to the problems of dwindling state aid and the city's growing budget (Swierz as quoted by Niquette 2012). Mayor Charles P. Sammarone noted that with the extra revenues generated from the sale and production of Ohio's oil and gas resources, the city could tear down many of its abandoned and blighted properties (Youngstown currently has more than 5,000 vacant structures). He also added that property owners within the city are already allowed to lease their lands for development (Niquette 2012).

City of Akron, Ohio

Akron is another older industrial city located in eastern Ohio. The city operates under a mayor-council form of government. Ten of the 13 council members represent districts, whereas the remaining three serve in an at-large capacity. The council promulgates ordinances, issues resolutions, levies taxes, and borrows money. An elected full-time mayor manages the city's day-to-day operations. The mayor may also veto council legislation, although the council may override with a two-thirds vote (City of Akron 2015a). In terms of the city's socio-economic demographics, its educational levels are close to state averages, although the city is home to considerably more renters and persons living below the poverty level (Table 7.11).

Table 7.11 Socio-Economic Characteristics of Akron

Characteristic	*Akron*	*Ohio*
Homeownership rate, 2009–2013	54.1%	67.5%
Housing units in multi-unit structures, percent, 2009–2013	29.3%	22.9%
Median value of owner-occupied housing units, 2009–2013	$83,900	$129,600
Per capita income in past 12 months (2013 dollars), 2009–2013	$19,968	$26,046
Median household income, 2009–2013	$33,909	$48,308
Persons below poverty level, percent, 2009–2013	27.5%	15.8%
Education, bachelor's degree or higher	20.5%	25.6%

Source: U.S. Census Bureau (2016)

The city issued a sustainability plan in 2009. Known as 'Greenprint for Akron,' the plan signaled the city's support for policies that protected the environment, reduced greenhouse gas emissions, improved quality of life, and grew the city's economy. As part of its plan, the city has conducted a greenhouse gas inventory, pursued green economic development, and sought outside smart growth funding (City of Akron 2015b).

The Akron City Council has supported and opposed state policy relative to oil and gas development. City leaders, for example, opposed House Bill 8 (HB 8), which would have permitted forced pooling of public parklands into a drilling unit without prior public permission. Mike Johnson, Metro Parks' interim director, noted that HB 8 "takes local control away" (Johnson as quoted by Warsmith 2015). Local leaders have also bemoaned the centralization of the oil- and gas-siting authority. At a 2015 city council meeting, Councilwoman at-large Linda Omobien explained that state policy has eliminated meaningful choice relative to oil and gas development. She noted that state policy has "left the city with no choice but to sell the oil leases, or risk losing the revenue to neighboring properties" (Omobien as quoted by Harper 2015). Omobien continued that should the city refuse to lease its lands for development, operators would access oil and gas via adjacent private property (Harper 2015).

Despite its unease about forced pooling, the city of Akron has benefited from oil and gas production:

- Since 2004, the Akron City Council has approved 33 wells, including one well located near a high school.
- In 2015, the city approved two additional wells.

(Harper 2015)

Table 7.12 Akron Support for Fracking

Year Leased	*Total Number of Leases*	*Acres Leased*	*City Revenues (in thousands)*
2004	1	4	305
2005	2	915	357
2006	2	20.5	361
2007	5	112.5	364
2008	4	977.5	500
2009	6	62	352
2010	4	73	483
2011	4	16	486
2012	2	36	Data unavailable
2013	3	38	Data unavailable
Totals	33	**2254.5**	

Source: Harper (2015)

- As of 2016, the city had more than 215 oil and gas wells on city lands; the wells have generated hundreds of thousands of dollars for the city.

(Harper 2015)

These Ohio communities show several factors related to local defiance, ranging from concern over the environment, the impact of a local disaster, and a belief that the state has abdicated its responsibilities to meaningfully regulate production (Table 7.13).

Table 7.13 Sources of Local Concern in Ohio

Factor	*Munroe Falls (defiance)*	*Broadview Heights (defiance)*	*Youngstown (alignment)*	*Akron (alignment)*
Negative impacts to environment	Yes	Yes	No evidence found	No evidence found
Lack of information sharing (emergency response)	No evidence found	No evidence found	No evidence found	No evidence found
Focusing events	No evidence found	Yes	No evidence found	No evidence found
Negative impacts to homeowners	No evidence found	Yes	No evidence found	Yes
Poor state oversight or policies	Yes	Yes	No evidence found	Yes

Source: Author generated

City of Arlington, Texas

The city of Arlington is located in the heart of the Dallas-Fort Worth Metroplex and is atop the resource-rich Barnett Shale. Its leadership team consists of an elected mayor and eight-member council (elected for two-year terms). Of the nine elected officials, the mayor and three council members represent the city at large. The remaining council members are elected via geographical district. Elected officials establish the city's priorities and pass legislation (including the city's oil and gas provisions) (City of Arlington N.d.). The city's administrative staff is led by an appointed city manager. Together, they oversee a large and diverse community, which is fairly representative of the state in terms of median home values, educational levels, and persons living below the poverty line, as shown in Table 7.14.

Table 7.14 Socio-Economic Characteristics of Arlington

Characteristic	*Arlington*	*Texas*
Population	383,204	
Homeownership rate, 2009–2013	57.4%	63.3%
Housing units in multi-unit structures, 2009–2013	35.4%	24.2%
Median value of owner-occupied housing units, 2009–2013	$130,000	$128,900
Per capita income in past 12 months (2013 dollars), 2009–2013	$25,456	$26,019
Persons below poverty level, 2009–2013	16.6%	17.6%
Education, bachelor's degree or higher, percentage of persons age 25+, 2009–2013	29.1	26.7%

Source: U.S. Census Bureau (2016a)

The city is home to 56 drill sites and 340 wells operated by seven different producers (City of Arlington 2014). In addition to relevant state rules, the operators producing within the city must comply with the provisions outlined in Municipal Ordinance 11–068 (Gas Drilling and Production). A staff of four, housed in the Community Development Department, oversees and enforces the city's policies (City of Arlington 2016a, 2014, 2011). Several noteworthy rules governing development are included, as follows:

- Operators must complete an application packet and plans.
- Operators must pay fees, insurance requirements, and bonding/surety provisions.
- Before beginning production, operators must receive a local special use and drilling permit along with TCEQ and RRC permits. Both city permits go through a public hearing process, which may take up to 12 weeks to complete.
- The ordinance authorizes the city to charge a road damage fee. The city can also collect replacement costs for road segments damaged by

industry (calculated by current cost per square yard of road surface material, installation, and labor).

- City policy establishes well setbacks from protected buildings and city parks at 600 feet. If the site is within the buffer area, operators must receive the property owner's consent before drilling. The council retains the authority to increase the setback distance.
- The city may require operators to install site features such as "fencing, screening, lighting, delivery times, noise levels, tank height, or any other matters reasonably required by public interest."
- The city retains the power to refuse and/or reject a well permit for any well located within the city's limits. The council may also accept, reject, or modify the drilling permit based on public health and safety.

Arlington has benefitted from oil and gas production primarily through the creation and maintenance of the Arlington Tomorrow Foundation (ATF), have used the fund to support various community development projects, including counseling services, pet micro-chipping, vaccinations, job training programs, and citywide concerts. In 2015, the ATF fund reached its statutory cap of $100 million, meaning that additional oil and gas revenues are scheduled to be transferred into the city's general, health insurance, and Arlington Public Finance Authority funds (Arlington Tomorrow Foundation 2015). In fiscal year (FY) 2016, the city collected $4.33 million in oil and gas revenues (up from $3.98 million in FY 2014) and earmarked them as follows:

- $1.3 million in technology upgrades for various city departments
- $215,000 for the library for e-content and collection development
- $2.17 million dedicated towards heavy fleet replacement
- $200,000 for capital improvements in the city's parks

(City of Arlington 2016)

City-collected fees are noted in Table 7.15.

Table 7.15 City of Arlington Oil and Gas Fees

	Actual FY 2014	*Budgeted FY 2015*	*Adopted FY 2016*
Drilling/gas well inspection fees	$551,000	$348,000	$261,000
Gas well re-inspection fee	$472,000	$578,000	$676,000
Gas well supplemental fee	$54,830	$30,000	$3,500
Gas well operational permit fee	0	$449,750	0
One-time general gas funds	$3,980,000		$4,030,868
Gas well response (expenditures)	$350,350	$452,392	$426,704

Source: City of Arlington (2016)

Arlington has experienced the darker side of development as well. On April 11, 2015, Vantage Energy spilled 42,800 gallons of fracking fluid onto city streets, sewers, and bodies of water. It also forced the evacuation of approximately 100 homes. After four attempts, the spill was controlled. City staff concluded that the spill released toxic chemicals, but did not cause any significant or permanent environmental damage. The city also reported a smaller spill of 1,500 gallons about a month earlier (Shipp 2015).

City leaders blamed Vantage and issued three citations and sought to be reimbursed for $84,000 in spill-related costs (Ciesco 2015). City Fire Chief Don Crowson described the operator's response as unacceptable, highlighting that Vantage Energy reported the spill to local emergency officials two hours after it began and that the call originated from Pennsylvania. The city's elected leaders have also questioned the delay in alerting local first responders and why the call came from an out-of-state location (Shipp 2015). Effects from the spill appeared to contribute to the Arlington City Council's decision to deny Vantage Energy a lease for a 12-acre site near Lake Arlington (Unger 2015).

City of Denton, Texas

Denton, Texas, home of the University of North Texas and to 125,000 fairly well-educated and affluent residents, is located at the northern edge of the Dallas-Fort Worth region. The city council is composed of four members representing specific districts, two at-large council members, and a mayor. Council members serve two-year terms, with a maximum of three consecutive terms. An appointed manager oversees the city's day-to-day operations. The city's environmental goals include improving regional air quality and mitigating climate change (City of Denton 2016). Socio-economic characteristics of Denton's residents is given in Table 7.16.

Table 7.16 Socio-Economic Characteristics of Denton

Characteristic	*Denton*	*Texas*
Population	128,205	
Housing units, 2010	46,211	9,977,436
Homeownership rate, 2009–2013	48.4%	63.3%
Housing units in multi-unit structures, 2009–2013	39.7%	24.2%
Median value of owner-occupied housing units, 2009–2013	$148,100	$128,900
Per capita income in past 12 months (2013 dollars), 2009–2013	$24,089	$26,019
Persons below poverty level, 2009–2013	20.6%	17.6%
Education, bachelor's degree or higher, percentage of persons age 25+, 2009–2013	37.6	26.7%

Source: U.S. Census Bureau (2016c)

The city's Gas Well Inspections (GWI) Division, part of the developmental services department, inspects and enforces policies for the city's 277 active wells, and for an additional 225 wells that are located on areas that the city considers part of its extraterritorial jurisdiction (ETJ). An additional 80 inactive wells exist within the city as well as 33 that are located in its ETJ (City of Denton 2016a). Before becoming the flashpoint for local activism within Texas, the city of Denton benefited from oil and gas production. Between 2004 and 2014, for example, oil and gas production contributed more than $30 million in local revenues.

Like defiance in other communities, Denton's activism was the culmination of a number of factors. Before the ban attempt in 2014, city leaders enacted a number of oil and gas restrictions:

- Created a task force in 2011 to study and consider potential oil and gas regulations
- Increased the required buffer distance between a drilling site and protected buildings from 1,000 feet to 1,200 feet
- Passed ordinance 2014–137, which established a fracking moratorium
- Passed ordinance 2014–276, which extended the fracking moratorium for an additional 45 days

Anti-fracking efforts were also aided by two localized focusing events. During the morning of April 19, 2013, Denton's emergency responders answered a call about a potential gas well blowout. The event forced the evacuation of several nearby homes, created a temporary no-fly zone because of the well's proximity to a nearby airport, and released 1,281 pounds of volatile organic compounds. The well was capped within several hours (it never ignited) and, according to the TCEQ, caused no permanent environmental damage (City of Denton 2016a). In May 2015, lightning ignited a well head and other equipment in the city's Westgate neighborhood. The blaze took firefighters four hours to contain. Neither event caused permanent or significant damage to adjacent areas (City of Denton 2016a, 2016b; Heinkel-Wolfe 2015a).

Throughout 2014 however, opposition continued. On July 15, 2014, Denton council members rejected a proposal to ban fracking (not drilling) with a 5–2 vote. Despite the loss at city hall, activists succeeded in placing the ban on the November 2014 ballot. It passed with 59 percent support (Dropkin and Henry 2015). Buchele (2014) surmised that concerns about public health and quality of life underscored the public's support for the ban. Cathy McMullen, the president of the Denton Drilling Awareness Group, commented that the (then proposed) local ban was the result of being out of options, "The city is allowing fracking to happen right in our backyards" (McMullen as quoted by Malewitz 2014).

State lawmakers preempted the Denton ban with HB 40 (NBC-DFW 2014). The city, in a 6–1 vote, repealed the ban and amended other sections of its oil and natural gas ordinance, with several significant changes noted here (Baker 2015; Heinkel-Wolfe 2015; Heinkel-Wolfe 2015b). The new oil and gas ordinance mandates that:

- Operators must still complete an application packet.
- Operators must still comply with the city's fencing, signage, safety, landscaping, dust, noise, lighting, and gating provisions.
- The city manager may still employ inspectors, who may enter and inspect oil and gas sites and facilities to ensure compliance with applicable laws.
- The city may also charge a road impact fee.
- Operators must still submit a water conservation plan, information about the use of erosion and sediment controls, a hazardous materials management plan, and emergency response information.
- A new setback distance of 1,000 feet (decreased from 200 feet) from protected buildings and parks.
- New buildings may not be within 250 feet of combustible sources on well sites.

The Denton 'story' has led to a number of relevant perspectives and insights into rights of local government, property owners, and state government:

- Denton Mayor Chris Watts described the impetus behind the city's efforts to ban additional wells within the city as related to the democratic rights of local governments. He added that it is important for the city to enforce "our ordinances . . . that have been voted upon by the community."

  (Watts as quoted by NBC-DFW 2014)

- Following the passage of Denton's fracking ban, the Chairperson of the Railroad Commission of Texas, Christi Craddick, stated, "It's my job to give permits, not Denton's. We're going to continue permitting up there because that's my job."

  (Craddick as quoted by Barnett 2014)

- The Program Director for the Texas Campaign for the Environment, Andrew Dobbs, suggested that "local governments are going to have a rude awakening . . . this is a power grab [by industry].

  (Dobbs as quoted by Baker 2015)

- Texas Oil and Gas Association President Todd Staples argued that HB 40 provides state and local lawmakers with clarity, consistency, and effective boundaries "more clearly than ever, [stipulated the] appropriate roles for the state and local governments in regulating the oil and natural gas industry."

(Staples as quoted by Heinkel-Wolfe 2015)

City of Fort Worth, Texas

The city of Fort Worth is also part of the larger Dallas-Fort Worth region and Barnett Shale play. It is led by nine council members and an appointed city manager. The council consists of eight members elected by district and a mayor who is elected as an at-large member. An appointed manager oversees the city's administrative operations for nearly 800,000 residents (City of Fort Worth 2016). Fort Worth is also committed to a three-pronged environmental plan: improving biking and walking transportation infrastructure, water conservation, and energy efficiency. The city's socio-economic characteristics are given in Table 7.17.

Table 7.17 Socio-Economic Characteristics of Fort Worth

Characteristic	*Fort Worth*	*Texas*
Population, 2014 estimate	812,238	
Homeownership rate, 2009–2013	58.5%	63.3%
Median value of owner-occupied housing units, 2009–2013	$120,700	$128,900
Per capita income in past 12 months (2013 dollars), 2009–2013	$24,489	$26,019
Median household income, 2009–2013 INC110213	$51,315	$51,900
Persons below poverty level, 2009–2013 PVY020213	19.3%	17.6%
Education, bachelor's degree or higher, percentage of persons age 25+, 2009–2013	26.5%	26.7%

Source: United States Census Bureau (2016e)

Economic output tied to production in Barnett Shale contributes millions of dollars to the North Texas economy and especially the city of Fort Worth. Economists are anticipating that between 2013 and 2023, local government above the Barnett Shale (including cities, counties, and school districts) are likely to top more than a billion of dollars' worth of economic output because of oil and gas (Fehlng 2014; Perryman Group 2014) (Tables 7.18 and 7.19).

The city's 1,976 producing wells are subject to rules governing well security, erosion and sediment control, signage and fencing, noise and

Table 7.18 Estimated Potential Impact of Oil and Gas Exploration, Drilling, and Operations; Royalty and Lease Payments; and Pipeline Construction, Maintenance, and Operations in the Barnett Shale over the Next 10 Years

*Core Counties**				
Economic Benefits (monetary values in billions of constant 2013 dollars)				
	Drilling	*Royalties*	*Pipeline*	*TOTAL*
Total expenditures	$324.1	$7.0	$3.2	$334.2
Gross product	$117.1	$3.5	$1.4	$122.1
Personal income	$68.2	$2.2	$1.0	$71.4
Retail sales	$27.7	$2.2	$0.4	$30.3
Employment (person-years)	1,012,181	46,431	17,895	1,076,507
Fiscal Benefits (monetary values in millions of constant 2013 dollars)				
Local governments	$4,618.7	$187.1	$65.7	$4,871.6

Source: Fort Worth Chamber of Commerce (2014)

odor, and timing standards (City of Fort Worth 2016a). The city also has a setback distance of 600 feet, which applies to residential, public, and religious buildings; parks; and hospitals (unless the operator acquires a waiver from the property owner and the city council) (Malewitz 2013). City policymakers established the 600-foot setback after commissioning an air-quality impact study in 2011. The study committee included Fort Worth citizens, local stakeholder groups, environmentalists, and representatives of the natural gas industry, and found minimal public health dangers for areas located outside of the 600-foot bubble.[1] The analysis included more than 150 different pollutants. Highlights of the study are included as follows:

1. "Ambient air monitoring data provide no evidence of 24-hour average concentrations reaching levels of health concern for acute exposures."

   (ERG 2011, 5.15.)

   - "Out of more than 15,000 air pollution measurements, only one exceeded a TCEQ health-based screening level for short-term exposures.

     (ERG 2011, 5.19.)

   - "For nearly every pollutant considered, the 600-foot setback distance appears to be adequately protective of public health, even for the highest-emitting sites."

     (ERG 2011, 5.36.)

Table 7.19 Impact of the Barnett Shale on Business Activity in Tarrant County

Estimated Current Annual Impact (2013) of Sources of Stimulus Associated with Barnett Shale

Sector	*Total Expenditures* (2013 dollars)	*Real Gross Product* (2013 dollars)	*Personal Income* (2013 dollars)	*Employment* (permanent jobs)
Agriculture	$78,485,707	$22,450,715	$15,016,599	233
Mining	$11,091,465,997	$2,431,779,927	$1,121,874,441	5,357
Construction	$1,548,781,830	$778,347,307	$641,406,644	8,794
Non-durable Manufacturing	$1,218,233,635	$358,222,181	$188,516,607	3,059
Durable manufacturing	$730,366,208	$285,687,923	$187,191,695	2,624
Transportation and utilities	$787,298,626	$353,025,957	$215,134,670	2,569
Information	$223,098,925	$137,289,971	$59,417,824	546
Wholesale trade	$499,766,034	$337,836,546	$194,799,530	2,120
Retail trade	$1,930,227,273	$1,440,086,954	$835,554,126	25,111
Finance, insurance, and real estate	$2,597,300,411	$795,051,035	$243,906,746	2,366
Business services	$566,670,291	$339,896,296	$277,268,277	3,281
Health services	$438,916,852	$306,895,800	$259,483,229	4,168
Other services	$786,585,084	$400,250,149	$319,904,047	7,172
TOTAL	$22,497,196,872	$7,986,820,763	$4,559,474,436	67,400

Note: Includes (1) drilling, exploration, and production; (2) royalty and lease payments; and (3) pipeline maintenance and operations.

Source: Fort Worth Chamber of Commerce (2014)

City of Mansfield, Texas

The city of Mansfield is an affluent and well-educated Fort Worth suburb, with high rates of home ownership and educational achievement, as shown in Table 7.20. Its leadership includes an elected city council and city manager. The mayor and council members are each elected at large and serve three-year terms (with no term limits) (City of Mansfield 2016).

As of 2015, the city's oil and natural gas inventory consisted of 57 approved well pads and 206 wells, which consumed approximately 350 acres (City of Mansfield 2015). Like many other communities, the city of Mansfield has codified many of its oil and gas standards (Chapter 114)

Table 7.20 Socio-Economic Characteristics of Mansfield

Characteristic	*Mansfield*	*Texas*
Homeownership rate, 2009–2013	80.2%	63.3%
Housing units in multi-unit structures, 2009–2013	12.3%	24.2%
Median value of owner-occupied housing units, 2009–2013	$179,600	$128,900
Households, 2009–2013	18,862	8,886,471
Persons per household, 2009–2013	3.03	2.82
Per capita income in past 12 months (2013 dollars), 2009–2013	$36,199	$26,019
Median household income, 2009–2013	$94,112	$51,900
Persons below poverty level, 2009–2013	6.4%	17.6%
Education, bachelor's degree or higher, percentage of persons age 25+, 2009–2013	41.4%	

Source: United States Census Bureau (2016i)

(City of Mansfield 2015a). Under city law, designated personnel may inspect sites (inspections are performed by the fire marshal and a well inspector). The city has also enacted site security, signage, and fencing standards. Operators must also receive a specific use permit (SUP) and a gas well permit before drilling (Table 7.21). These permits can be concurrently processed and may take up to six months (City of Mansfield N.d.).

As HB 40 was working its way through the statehouse, Mansfield City Council members were also considering new oil and gas regulations, including an increase to the city's setback distance (Peterson 2015).

Table 7.21 Mansfield Permitting Process

Step 1	Review and approval of the SUP takes approximately three months once the city has received the application. Through the SUP request, the city determines the drill site's size and the legally allowed number of wells on the site.
Step 2	The city's Planning and Zoning Commission then holds a public hearing relative to the SUP request. It also makes a recommendation to the city council as to whether to approve the permit. Once at the council stage, the SUP receives two additional public hearings and then city council members vote to approve or reject the SUP. The SUP is valid for five years.
Step 3	Once city officials approve the SUP, the operator must file a gas-well permit request before conducting any drilling activities. The process typically takes between one and two months after the city receives the submission. City council does not need to review gas well permits if they are issued in accordance with the already approved SUP.

Source: City of Mansfield (2015a)

Industry and its allies, such as the Citizens for Mansfield group, argued that an increase in the city's setback distance would hurt the city's economy and overly burden industry. Conversely, grassroots opponents sought to expand the city's buffer space from 600 to 1,500 feet (Allen 2013). Council members amended the city's regulations, but opted to keep the setback distance at 600 feet (Peterson 2015). City Attorney Allen Taylor explained the decision to maintain the setback distance as "push too far you begin to invade the rights on the other side of the equation, which could put the city at legal risk" (Taylor as quoted by Cadwallader 2015). Other changes included the following:

- Operators are required to meet with fire department officials annually so that emergency response plans may be reviewed.
- Fracking is banned on Sundays and holidays recognized by city hall.
- Operators must conduct pre- and post-drilling water tests for any water well within 500 feet of a gas well (if requested by the owner of the well).
- City permits expire after five years.
- Operators' equipment must be electric and connected to Mansfield's power grid.
- Ordinance requires a 1,000-foot setback between new wells and hospitals, jails, and nursing homes.
- The city may monitor wells and test for harmful air emissions.

(Cadwallader 2015)

As shown in Table 7.22, the local case studies from Texas also reveal several factors related to local defiance, including focusing events and home values.

Table 7.22 Sources of Local Concern in Texas

Factor	*Denton*	*Arlington*	*Fort Worth*	*Mansfield*
Negative impacts to the environment	Yes	No evidence found	No evidence found	No evidence found
Lack of information sharing (emergency response)	No evidence found	Yes	No evidence found	Yes
Focusing events	Yes	Yes	No evidence found	No evidence found
Negative impacts to homeowners	Yes	No evidence found	No evidence found	Yes
Poor state oversight or policies	No evidence found	No evidence found	No evidence found	No evidence found

Explanations

Although the cities represent three different states, the case studies reveal a number of commonalities in terms of the local experience. The emergence of such factors suggests that concerns related to environmental protection, impacts to homeowners, emergency management, and focusing events often underpin challenges to the state. Importantly, by identifying these factors early on, stakeholders can begin the difficult conversations that are needed before identifying potential governing solutions.

It is clear that cities are not a unified block for or against oil and gas development or fracking as an extractive process. Voters in Youngstown, for example, rejected multiple fracking bans. In Youngstown, Ohio; Fort Worth, Texas; and Greeley, Colorado, leaders have praised the industry for its ability to create jobs and generate investment within their communities. Ohio's communities, like those in Colorado and Texas, have also leased their lands and water resources to support shale development (Downing 2013; see also Gorman 2013). Even cities that have blocked development have benefited from oil and gas operations.

It is also clear from the case studies that cities are willing to take steps to mitigate potential risks to citizens and that they are cognizant of their legal risks and limitations. Before the passage of Denton's fracking ban, Mayor Mark Burroughs summarized the risks associated with local defiance: "If it [the ordinance] does pass, the city has to follow it . . . but we could be bound to enforce an illegal act, which throws into a whole panoply of open issues" (Burroughs as quoted by Dropkin 2014). Leaders in Loveland and Fort Collins, Colorado, expressed similar worries about a state lawsuit.

Focusing Events

The case studies shed light on the important role that focusing events likely play in explaining why cities defy their state government. Whether it is spills, concerns over emergency management, the novelty of development, or environmental damage, fracking is on the minds of many. In the cities of Longmont, Denton, and Arlington, for example, experiences with oil and gas production illustrate the interplay between local defiance and focusing events. Defiance in the city of Longmont began in earnest after residents pressed the city over leaks near the Rider Well. Located near a middle school and several homes, concern over the site became a focal point for advocates of more restrictive oil and gas policies. Concerns over potential spills also underpinned municipal activism in Broomfield and Fort Collins (Rochat 2012, 2012a; Quinn 2013). Localized focusing events appear to have spurred action in Denton, Mansfield, and Arlington as well. In each community, groups took advantage of the event to push for additional municipal oversight. This finding is consistent with the policy literature that has found that the political impacts of a focusing event are often amplified when stakeholders use events to their political advantage (Birkland and Nash 2000). Birkland (2011), detailed that following the shock, stakeholder groups pursue a variety of strategies such as introducing their preferred solution to policymakers, highlighting the gaps in current regulation, gaining additional media attention, and reframing the issue to their preferred terms. They often do so, according to Birkland (2011), to challenge and change the status quo.

Why would a threat precipitate citizen action and mobilization? Miller and Krosnick's work (2004) offers some clues as to the role that real and perceived 'threat' plays relative to motivation and decision-making. Citizens, when they perceive something as a threat, become more motivated and look for plausible ways to mitigate the threat. For activists in Denton, Arlington, and Mansfield, a reasonable strategy following a focusing event may be to work with city leadership and collect additional information to highlight regulatory gaps, and to reframe the oil and gas question.

Breakdowns in Trust

Data revealed some local defiance was related to a breakdown in trust between city officials and the state agencies, such as the ODNR (Fisk 2016; Henry 2013b). Many of the reasons listed in Munroe Falls' anti-fracking resolution were explicitly tied to the belief among some city officials that the ODNR failed to adequately protect public health and the environment. Paul Marnecheck, North Royalton Ward 4 council member, exemplified this sentiment argued, "North Royalton can do a better job with how to have this in our city than a bureaucrat in Columbus." He continued that when cities are vested with strong home rule powers, they are better able to protect the unique character of their communities and to better reflect the will of their constituents (Marnecheck as quoted by Anton 2013). City of North Royalton Ward 6 Councilman Dan Kasaris observed that local governments already "regulate the placement of fences, driveways, sheds, and houses but not an oil well . . . to me that fact is inconsistent with the rule of those who reside within the boundary of any city" (Kasaris as quoted by Anton 2013). Brunswick City Council member Brian Ousley was quoted as saying "he believed that the state should not tell us [the city] what to do." He later added he is not "business unfriendly, but I am resident friendly . . . threats by businesses cannot bully our citizens" (Ousley as quoted by Lisik 2013).

Note

1 The group also recommended that the city implement reasonable precautions and install vapor recovery units on storage tanks, three-way catalysts, catalytic oxidizers on compressor stations, and adequate fencing. Other recommendations called for the increased use of electric compressor engines and low-bleed or no-bleed pneumatic valve controllers. Finally, the report called for improved inspection and maintenance of industry equipment (City of Fort Worth 2011).

8 Local Oil and Gas Trends

In each of the three states examined in preceding chapters, a small number of the sampled communities have challenged their state's preemptive authority. Such decisions may be influenced by a variety of factors, including city-level socio-economic characteristics, the values of city hall, voter turnout, and the city's overall commitment to pursuing sustainable economic growth. Some of these variables can be evaluated statistically.

Sampling Process

To examine possible antecedents of local defiance, this chapter again relies on a purposive sampling strategy. Like the preceding chapter, the risks and rewards of oil and natural gas production do not fall along city, county, or state lines. Deposits can also be unevenly distributed within local jurisdictions. With this in mind, the first filter limits the analysis to only those cities located in counties with active oil and natural gas exploration/production. Where possible, emphasis is given to communities located near significant population centers. Because oil and gas production is located either within or near these communities, residents and leaders are more likely to experience the costs and benefits of development directly (Rabe and Borick 2013; Jacquet 2014, 2012; Opsal and O'Connor Shelley 2014; Wiseman 2009). To keep the distribution of sampled cities approximately equal across the three states, this process produced a sample of 60 Ohio communities, 60 Texas communities (mostly in the highly populated Barnett Shale region), and 48 in Colorado.

Data were collected in the following process:

1. For each of the sampled communities, key word search was performed in each municipality's website and codebook (if necessary) using the following search terms/phrases: fracking, hydraulic fracturing, drilling, moratoria, bans, community rights, oil and gas, zoning, and local authority.

2. If this first step failed to locate data relative to the dependent variable, a LexisNexis and Google News search was then performed that linked the city of interest with the search terms.
3. Findings were triangulated with newspaper articles, press releases, and whitepapers, if available.

Dependent Variable

The dependent variable is a measure of municipal policy actions in response to oil and gas development. Local authority includes the ability to enact and enforce land use plans, promulgate ordinances, and engage in policy actions that safeguard public health and safety. Cities may also lobby state officials, hold press conferences, stage protests, enter into negotiations with other stakeholders, and release letters and amicus briefs (Baumgartner and Jones 1991; Berman 2003; Bruhl, Linder, and Sexton 2013; Kincaid 1999; Pralle 2006; Riverstone-Newell 2012; Sherman 2011; Zimmerman 1995, 2012). Each of the sampled cities' policies towards oil and gas development and/or fracking was organized on a continuum ranging from actions that encourage development to actions that impede or preclude development (Figure 8.1).

Municipal Governments' Oil and Gas Policy Action Scale

States have traditionally limited municipal involvement. Colorado, for example, allows city and county governments to call for hearings, issue special-use permits, and mitigate road impacts (COGCC 2008). Ohio law and jurisprudence enables townships to enter into road maintenance agreements but curtails local authority to promulgate ordinances specific to oil and gas development. Texas gives local governments a greater role in well-siting decisions—although HB 40 has restricted local siting authority and commits local regulation to a commercially reasonable standard (Baker 2015).

<table>
<tr><th colspan="6">Opposition</th></tr>
<tr><td>Bans and/or moratoria</td><td colspan="2">Zoning regulations that exceed the state standards and limit where development may occur</td><td colspan="2">Banned on municipal-owned property</td><td>Resolutions against the practice</td></tr>
<tr><th colspan="2">Working within the State Framework</th><th colspan="4">Support</th></tr>
<tr><td colspan="2">Voluntary agreements with industry that still permit development</td><td colspan="2">No action, special-use permits, or resolution in support of development</td><td colspan="2">Ancillary actions (i.e., selling/leasing excess water)</td></tr>
</table>

Figure 8.1 Local Options

Source: Author generated

Table 8.1 Regulatory Options for Municipalities in Colorado, Ohio, and Texas

Policy Responses	*Coding Value*	*Local Defiance in Colorado*	*Local Defiance in Ohio*	*Local Defiance in Texas*
Bans, moratoria, oil/gas ordinances*	0	Yes*	Yes*	Yes*
Zoning regulations*	1	Yes*	Yes*; treated as local defiance by the Ohio oil and gas industry	No
Bans on municipal property	2	No	No	No
Voluntary agreements	3	No	No	No
Resolutions for local control/ anti-fracking	4	No	No	No
No action/resolutions in favor/ special-use permits that do not conflict with state law	5	No	No	No
Actions increasing development	6	No	No	No

*Actions that have prompted lawsuits or state legislation

Source: Author generated

Despite the financial and legal risks, some municipal, county, and township governments have sought more control over oil and gas operations taking place within their cities. Communities have enacted zoning plans that limit the locations of development, mitigate site nuisances, regulate traffic routes and times of deliveries, and assess impact fees on developers. When legal options fail or are unavailable, cities have also utilized more symbolic measures, including protest tactics, press releases, and passing proclamations, and have authored amicus briefs to advance their legal claims (Baumgartner and Jones 1991; Berman 2003; Bruhl, Linder, and Sexton 2013; Fisk 2016; Kincaid 1999; Pralle 2006; Riverstone-Newell 2012; Sherman 2011; Zimmerman 1995, 2012).

Bans and moratoria are legal maneuvers that block development. Cities may also rely on ordinance(s) or zoning plans that place specific standards on oil and gas operations within the community. In other cases, local regulations may restrict the use of horizontal drilling but still permit vertical drilling. Local regulations, like bans, carry legal risks, especially if they include setbacks or other technical restrictions that exceed state standards. Moving slightly towards the center are zoning and land use plans that narrow the locations available to oil and gas operations (i.e., allows development only in zones designed for oil and gas exploration). Finally, cities may restrict or ban operators from drilling on public property.

These are also the policy options that are most likely to elicit a fight with industry and industry-friendly state lawmakers (Fisk 2016).

Closer to the continuum's midpoint are resolutions against fracking or state's oversight of drilling and voluntary agreements. These actions remove many of the legal risks associated with local defiance because they do not preclude development (Barnes 2013; Fisk 2016). Voluntary agreements seek to mitigate impacts to the city's air and water quality, siting issues, emergency management, disaster response, inclusion of public safety items (such as fencing), and information sharing (truck routes, chemicals) with first responders.

The last two 'items' on the continuum demonstrate support for oil and gas development and are also within the state's regulatory and legal frameworks. Municipal governments may elect to work within the state's oil and natural gas policy framework and allow it to guide development. Finally, communities may lease their public spaces or water to operators (Barnes 2013; Fisk 2016).

Communities have options when oil and natural gas firms arrive at city hall. They might refuse, ignore, and impede operations. They may also opt to work cooperatively with state-elected and regulatory officials. However, when they do decide to pursue oppositional strategies, they take on substantial political and financial risk. In multiple states, courts have become involved—including Colorado, Ohio, New York, Louisiana, New Mexico, and Pennsylvania. Importantly, only Pennsylvania and New York courts have upheld local regulatory authority. In the others, courts have struck down local actions. And, in other states, lawmakers have added new statutory language that expressly restricts local autonomy.

Independent Variables and Statistical Technique

Because communities have a variety of actions available to them that constitute defiance, the first model measured both symbolic and legal local defiance, whereas the second narrowed defiance to only those communities that have utilized their legal authority to restrict oil and gas regulations. Logistic regression was applied to measures of defiance.

Policy researchers devoted to state oil and gas policies have linked opposition to oil and gas production to institutions and individuals who prioritize environmental protection (Boudet et al. 2014; Fisk 2016; Davis and Fisk 2014; Davis 2012). Although an imperfect measure, the presence and concentration of green jobs have been used to demonstrate municipal commitment towards environmental protection. Portney (2009), for example, described that one key way cities become more sustainable is by recruiting and retaining firms that offer 'green' jobs (see also Opp and Saunders 2013). Later work by Svara, Watt, and Jang (2013) also connected the presence of 'green' jobs to greater levels of municipal sustainability. They noted that cities that prioritize attracting green firms/jobs are often more sustainable than local governments that do not attach the same importance to sustainable economic development. Yi (2013) also supports the relationship between environmental protection and the presence of green jobs.

Local defiance in response to oil and gas development may be related to a city's commitment to sustainability (Portney 2009; Selman 2004; Svara, Watt, and Jang 2013; Yi 2013). The variable approximating commitment to sustainable economic development was standardized by dividing the number of green jobs (38 of the 39 Brookings' Institution green job estimates, with nuclear power jobs dropped) with the county's total number of non-seasonal jobs (as listed by the Bureau of Labor Statistics). The top half of cities were labeled high-sustainable economic development while the bottom half were considered low-sustainable economic development (Bureau of Labor Statistics, 2013a, 2013b).

Anti-fracking activists have utilized citizen-led ballot initiatives to restrict unconventional oil and gas drilling. This potential relationship was approximated through the most recently available county-level voter turnout data: Ohio, 2012 presidential election; Colorado, active registered voters as of 2014; and Texas, 2010 mid-term elections (Baumgartner and Jones 1991; Jones and Baumgartner 2012; Kingdon 1995). The bottom half of cities were labeled low-turnout green, whereas the top half were considered high turnout.

The two measures (green × turnout) were then combined to create an interactive green voter mobilization variable: 0 = Low green/Low voting area, 1 = Low green/High voting area, 2 = High green/Low voting area, and 3 = High green/High voting area and treated as a continuous variable:

- H1: Communities with a greater number of green jobs (as defined by Brookings Institution) will be associated with local defiance.
- H2: Communities with a greater number of voters will be associated with local defiance.
- H2a: Communities with a greater number of green jobs AND voters will be associated with local defiance.

Work by DiPasquale and Glaeser (1999) identified that home ownership influences behavior, including the willingness of citizens to solve local problems. Prior environmental policy research also has identified a relationship between greater wealth and a willingness to support environmental protection (Daley and Garand 2005; Howell and Laska 1992; Jones and Dunlap 1992; Matisoff 2008; Wood 2011). Socio-economic status was operationalized through a number of variables such as home values (natural log), education levels, percentage of owner-occupied homes, and per-capita income (natural log) (Table 8.2).

- H3: Communities with a higher socio-economic status will be associated with local defiance.

H4: Communities with higher home values will be associated with local defiance.
H5: Communities with higher rates of owner-occupied homes will be associated with local defiance.

Table 8.2 Exploratory Defiance Variables

Independent Variable	*Operationalization*	*Source*
Sustainable economic development	Green jobs—dichotomous high/low	Brookings Institution green jobs data
Voting turnout	Voter turnout—Dichotomous high/low	State Secretary of State pages
Wealth/affluence	Median household income (natural log) dropped based on concerns of multi-collinearity	2008–2012 ACS survey
Median home value	Median home value (natural log)	2008–2012 ACS survey
City/town education levels	The percentage of citizens who hold 'bachelor's degree or higher' out of the total number of adults in the community (dropped based on concerns of multi-collinearity)	2008–2012 ACS survey
Homeownership rate	The percentage of citizens who are homeowners (rather than renters) and who hold 'bachelor's degree or higher' out of the total number of residences in the community	2008–2012 ACS survey

The year 2008 was selected because it is often seen as the beginning of the fracking renaissance (Davis 2012).

Source: Colorado Secretary of State 2014; Texas Secretary of State 2014

A Sample of Colorado's Municipalities

Colorado's intergovernmental landscape has examples of conflict and collaboration (Eaton 2013; Lynn 2012). In some communities, city leaders/citizens have urged outright defiance to the COGCC's rules and are willing to challenge the state's authority. In other cities, city leaders and residents have been willing to negotiate with state officials and industry. Typically, these agreements exchange expedited permit times for more stringent performance standards (cities of Loveland and Erie). Finally, other Colorado cities are taking additional steps to encourage oil and natural gas development (Greeley, Grand Junction, and Aurora). An elected state official described Colorado's episodes of local defiance as:

Table 8.3 Aggregated Municipal Responses to Oil and Gas Development

Policy Responses (policy responses code)	*Frequency*	*Percent*	*Valid Percent*	*Cumulative Percent*
Bans or moratoria (0)	**4**	**8.3**	**8.3**	**8.3**
Zoning regulations (1)	**2**	**4.2**	**4.2**	**12.5**
Bans on municipal property (2)	2	**4.2**	**4.2**	**16.7**
Voluntary agreements (3)	4	8.3	8.3	25.0
Resolutions for local control/ anti-fracking (4)	4	8.3	8.3	33.3
No action/resolutions in favor/ special-use permits that do not conflict with state law (5)	24	50.0	50.0	83.3
Actions increasing development (6)	8	16.7	16.7	100.0
Total	48	100.0	100.0	

Data collected from municipal websites, codes, and news articles

Bold type signifies legal challenges

Source: Author generated

> "The overall state of state-municipal relations concerning fracking depends, at least in part, on the particular municipal jurisdiction. However, there is definitely a tension between the state and local jurisdictions, and in Fort Collins many residents have definite land-use, public health and welfare concerns. Such concerns were manifested with the passage of a local ballot measure to implement a 5-year moratorium on fracking within city limits."

A quick scan of municipalities shows that Colorado's communities have responded to fracking in a variety of ways.

Table 8.3 offers a snapshot of relationships in Colorado. The mean score, 4.21, is suggestive of a fairly cooperative relationship between cities and the state and one that typically sees local governments working within the COGCC's regulatory processes. Yet, bans and moratoria, especially among High green/High turnout communities have led to oppositional state-local relationships. These challenges have received the majority of the headlines and have generated high levels of attention from industry and state leaders.

Half of the sampled cities issued special/specific-use permits or conditional-use grants that impose some restrictions on operators but fit within state policies. In addition to working within COGCC rules, another subset of cities has also leased their public spaces and/or supplied excess water for oil and gas production, including the larger cities of Aurora (population: 340,000) and Greeley (population: 95,000) (Finley 2012).

The sample includes examples of local defiance. A third of the sample's local governments opposed (in some form) some aspect of the COGCC's regulatory approach. Several communities banned or blocked development and/or fracking through the enactment of bans, moratoria, or restrictive oil and gas ordinances, or through land use and zoning authority (before the State Supreme Court case). An additional 10 percent of the selected communities published proclamations and city council resolutions, issued press releases, and authored letters that were critical of the state's actions towards oil and gas production and fracking.

Colorado Statistical Results

The broader group of defiance produced two statistically significant relationships: the natural log of the community's median home value ($p = .05$) and cities committed to more sustainable forms of economic development ($p = .01$). The measure of voter turnout failed to reach statistical significance.[1] The results are presented in Table 8.4.

In this broader model of municipal defiance, each one unit increase (going from a 0 to a 1, etc.) in the city's sustainable development score was tied to a 90 percent decrease in the odds of a policy that conformed to the COGCC's regulatory framework, while holding the natural log of the city's median home value constant (Table 8.5). This is a dramatic decrease, but some of this effect is due to how 'greenness' was measured. For some cities, green jobs comprised 3 to 4 percent of the city's overall employment base. This means that to increase the city's green job

Table 8.4 Municipal Challenges to the State (broad)

	B	*S.E.*	*Wald*	*Df*	*Sig.*	*Exp(B)*
Natural log median home value*	–1.943	.990	3.850	1	.050	.143
Sustainable economic development*	–2.295	.905	6.433	1	.011	.101
Constant	26.130	12.270	4.536	1	.033	2.229E11

Variable(s) entered: Educational attainment, voter turnout, green/sustainable economic development interaction, percentage of owner-occupied homes, natural log of per capita income (dropped in favor of median home value [natural log]).

*Significant at the .05 level

Source: Author generated

percentage, it would need to attract green jobs at a substantially faster pace than it attracts overall jobs.

The natural log median home value also reached statistical significance. For every one-unit increase in the natural log median home value, the likelihood of the city's policy deferring to the COGCC decreased by 86 percent, holding all other variables constant. This number may also seem like a dramatic impact, but it is also related to how it was measured. A one-unit increase in the natural log median home value is the equivalent of a local community's median home value increasing from $100,000 to $272,000.

Table 8.5 Municipal Fracking Policies That Legally Challenge the State's Authority

	B	*S.E.*	*Wald*	*Df*	*Sig.*	*Exp(B)*
Sustainable economic development**	–2.341	1.267	3.416	1	.065	.096
Turnout**	–2.176	1.236	3.097	1	.078	.114
Constant	5.689	1.864	9.312	1	.002	295.523
Variable(s) entered: Educational attainment, voter turnout, green/sustainable economic development interaction, percentage of owner-occupied homes, natural log of per capita income (dropped in favor of median home value [natural log]).						

**Significant at the .1 level

Source: Author generated

The second model specified a narrower 'version' of municipal defiance and only counted communities whose policies represented a legal challenge to the state. Here, no variables reached the .05 level of statistical significance. However, if a .1 threshold is utilized, one variable is significant (sustainable economic development) and another (turnout) nearly reached statistically significant levels. For legal challenges to the state's preemptive authority, as a municipality goes from low to high in terms of its sustainability score, there was a 91 percent decrease in the odds that the city's policy fits within the state's regulatory framework, holding other variables constant. In a similar manner, a one-unit increase in voter turnout decreased the odds of enacting a municipal drilling policy that can be harmonized with state oil and gas goals/policies by 89 percent, while holding the green measure constant. Unlike the previous model, the natural log of a community's median home value failed to reach statistical significance. The remaining variables were not significant in any model.

A Sample of Ohio's Municipalities

The boom has contributed to examples of intergovernmental cooperation, opposition, and uncertainty in Ohio as well (DeWitt 2013; Fry

2013). The communities of Cincinnati, Munroe Falls, Athens, Broadview Heights, Oberlin, Mansfield, and Yellow Springs have limited the operations of oil and gas wells have impeded the installation of other industry infrastructure, or have attempted to restrict wastewater injection wells (Smith 2015). Other communities are more likely to view the sight of oil and gas wells as a welcome harbinger of new jobs and economic growth. Voters in the communities of Youngstown, Gates Mills, and Kent, for example, have rejected citizen-led pushes for fracking bans. Supportive local governments have also facilitated development by leasing their public spaces and supplying water to drillers. Before it banned future oil and gas wells, the city of Broadway Heights even received lease payments from oil and gas operators (Spector 2012).

Ohio's municipalities have several options available should they seek to prevent or regulate oil and gas development. Table 8.6 shows the relative frequency of each option. The sample data show that Ohio's communities vary significantly in their decision whether to challenge the policies of the Ohio Department of Natural Resources (ODNR). Most frequently, communities opted to work within the ODNR's framework and regulate production through the issuance of special-use permits (22/60). Other local governments responded to the boom by leasing city lands or water to industry. Together, these two non-oppositional segments accounted for a majority of the sample. Yet, local defiance is also taking place. A number of the sampled communities have challenged the state through symbolic

Table 8.6 Aggregated Municipal Responses to Oil and Gas Development in Ohio

Policy Responses (policy responses code)	*Frequency*	*Percent*	*Valid Percent*	*Cumulative Percent*
Bans or moratoria (0)*	7	11.7	11.7	11.7
Zoning regulations (1)^	15	25.0	25.0	36.7
Bans on municipal property (2)^	0	0	0	0
Voluntary agreements (3)^	0	0	0	0
Resolutions for local control/ anti-fracking (4)^	1	1.7	1.7	38.3
No action/resolutions in favor/ use permits that do not conflict with state law (5)	31	51.7	51.7	90.0
Actions increasing development (6)	6	10.0	10.0	100.0
Total	60	100.0	100.0	100.0

Data collected from municipal websites, codes, and news articles
^Symbolic challenges
*Substantive challenges

Source: Author generated

measures (i.e., issued a city council resolution). Another sizable group of cities restrict development through bans and other legal restrictions.

Ohio Statistical Results

A variety of factors may be associated with conflicted intergovernmental relationships in Ohio. Representatives of the city of Munroe Falls explained that their city's fracking restrictions were an exercise of its home rule authority rather than a referendum on fracking (Downing 2014). Smaller towns such as Oberlin, Athens, and Bowling Green have also advocated for more oversight and have raised alarms over the industry's influence at the statehouse (Brumfield 2013; Henry 2013b). Members of the Cincinnati City Council banned injection wells (legal as a zoning use), citing the need to protect citizens from health risks (Downing 2012). City Council Member Laure Quinlivan commented that the city's ban sought to "to make Cincinnati cleaner, greener, and smarter" (Quinlivan as quoted Downing 2012). Bruce Whitteberry, the assistant superintendent of Cincinnati Water Works (CWW) stated that the city's "goal is to know what is in a spill—and know about it ahead of time . . . that gives us the best shot to make sure all of our customers stay protected" (Smith 2014).

To measure the antecedents of local defiance, municipal responses were recoded into two dichotomous dependent variables. The broader assessment of local defiance produced two statistically significant relationships: sustainable economic development ($p = .035$) and cities located in counties with a greater concentration of voters ($p = .035$), as shown in Table 8.7. A second model dropped turnout and sustainable economic development as individual variables in favor of an interactive variable (green × turnout), but this did not produce any statistically significant relationships.

In this model, the sustainable economic development and turnout variables reported similar effects on the municipal policies. A one-unit increase in turnout (going from low to high in the number of active

Table 8.7 Municipal Challenges to the State (broad)

Variable	*B*	*S.E.*	*Wald*	*Df*	*Sig.*	*Exp(B)*
Sustainable economic development	–1.196	.567	4.445	1	.035	.302
Turnout	–1.196	.567	4.445	1	.035	.302
Constant	1.352	.528	6.549	1	.010	3.865

Variable(s) entered: Educational attainment, voter turnout, green/sustainable economic development interaction, percentage of owner-occupied homes, natural log of per capita income (dropped in favor of median home value [natural log]).

Source: Author generated

voters) was associated with a 70 percent decline in the odds that the local response action fits within the ODNR's regulatory framework, while holding the sustainable economic development score constant. Conversely, when turnout was held constant, a one-unit increase in the commitment to sustainable economic development score reduced the odds of a policy that defers to the state by 70 percent.

Table 8.8 Municipal Challenges to the State (narrow)

	B	*S.E.*	*Wald*	*Df*	*Sig.*	*Exp(B)*
Owner-occupied	−7.544	3.119	5.849	1	.016	.001
Constant	6.530	2.296	8.089	1	.004	685.43

Variables included educational attainment, voter turnout, green/sustainable economic development city, percentage of owner-occupied homes, natural log of per capita income (dropped in favor of median home value [natural log]).

Source: Author generated

In the narrower model of what constitutes local defiance, both the sustainable economic development and turnout measures failed to maintain their statistical significance. The interactive variable (green × turnout) also failed to reach significance. Interestingly, the percentage of owner-occupied homes became significant ($p = .016$), as shown in Table 8.8. For each one-unit increase in the percentage of a city's owner-occupied homes, there was a 99 percent decrease in the odds that the community's policy complies with the ODNR's regulatory scheme. The impact of owner-occupied homes appears fairly dramatic; however, much of this relationship's strength is due to how the variable is operationalized. In Youngstown, Ohio, for example, approximately 16,000 out of a possible 26,800 homes are owner-occupied. A one-unit change in this variable as applied to Youngstown would be equivalent to the homeowner rate increasing from .597 to 1.597. This rate is not possible (i.e., there cannot be more than 100 percent of homes)

A Sample of Texas's Municipalities

Texas's cities have responded to oil and natural gas exploration in a variety of ways. The city of Dallas, for example, sold mineral rights to Trinity East Energy for several million dollars in 2008 (Henry 2013). Several years later, city lawmakers passed a 1,500-foot setback, which, according to industry officials, blocks most development. Denton, Texas, the state's hotbed of local activism, has also enjoyed the fruits of fracking's finances. Between 2004 and 2014, industry remitted more than $30 million in lease payments and other revenues to the city (Schmall 2014).

Table 8.9 presents a small sample (N = 60) of municipal regulatory responses organized by a dichotomous measure of a high or low green city and high or low voter turnout percentages. The average policy score for the sampled communities was 3.425, suggesting some degree of compliance with RRC's regulatory framework. Table 8.9 also displays the relative frequency of each policy response.

Table 8.9 Aggregated Municipal Responses to Oil and Gas Development

Policy Responses (policy responses code)	*Frequency*	*Percent*	*Valid Percent*	*Cumulative Percent*
Bans or moratoria (0)*	3	4.8	5.0	5.0
Zoning regulations (1)*	14	22.6	23.3	28.3
Bans on municipal property (2)	1	1.6	1.7	30.0
Voluntary agreements exceeding the state (3)^	0	0	0	30.0
Resolutions for local control/ anti-fracking (4)^	10	16.1	16.7	46.7
No action/resolutions in favor/ special-use permits that do not conflict with state law (5)	22	35.5	36.7	83.3
Actions increasing development (6)	10	16.1	16.7	100.0
Total	60		100.0	
Data collected from municipal websites, codes, and news articles				

^*Symbolic challenges*

*Legal challenges

Source: Author generated

Several trends in Table 8.9 are worth highlighting. First, the most frequent policy response were actions that fit within the RRC's approach—the requirement of a special/specific-use permit and leasing water or selling mineral rights with no additional restrictions on operators. Even after HB 40, Texas grants municipal governments a greater degree of regulatory latitude when it comes to local oil and natural gas siting and management. Roughly a third of cities in the sample had zoning and other land use policies (zoning) that addressed fracking's surface-level impacts. Additionally, seven cities have well setback distances of greater than 1,200 feet, which industry considers de facto bans (data from author-generated table).

Texas Statistical Results

The first dependent variable defined defiance more broadly and includes bans, zoning policies, proclamations, resolutions, amicus briefs, or other maneuvers that symbolically record the city's opposition to state policy. The second dependent variable was limited to cities with policies that operationally conflict with state regulations.

The first model reported only one significant relationship, the natural log of the city's median home value (p = .008) (Table 8.10). For each one-unit increase in the natural log of the median home value, the likelihood of deferring to the RRC's regulations decreased by approximately 84 percent. However, because of how home values were measured, a one-unit increase is significant and would be the equivalent to jumping from a $100,000 median home to one valued at $272,000, assuming all other variables remained constant.

The second defiance model led to a similar result. A one-unit change was the equivalent to an 87 percent decline in the odds, while holding all other variables constant. It should be noted that the same 'real world' measurement limitation applies.

Table 8.10 Municipal Challenges to the State (broad)

	B	*S.E.*	*Wald*	*Df*	*Sig.*	*Exp(B)*
Median home value (natural log)	−1.825	.690	6.994	1	.008	.161
Constant	21.846	8.109	7.258	1	.007	3.073E9

Variable(s) entered: Educational attainment, voter turnout, green/sustainable economic development interaction, percentage of owner-occupied homes, natural log of per capita income (dropped in favor of median home value [natural log]).

Source: Author generated

Table 8.11 Municipal Challenges to the State (narrow)

	B	*S.E.*	*Wald*	*Df*	*Sig.*	*Exp(B)*
Median home value (natural log)	−1.984	.898	4.887	1	.027	.137
Constant	25.521	10.761	5.625	1	.018	1.212E11

Variable(s) entered: Educational attainment, voter turnout, green/sustainable economic development interaction, percentage of owner-occupied homes, natural log of per capita income (dropped in favor of median home value [natural log]).

Source: Author generated

Explanations

Data point to a city's inclination towards environmental protection, median home values, and home ownership as the factors associated with local defiance in Colorado, Ohio, and Texas. In Table 8.12, the summarized results are shown.

Table 8.12 Summary of Findings

Variable	*Symbolic and Substantive Defiance*			*Substantive Defiance*		
	CO	*OH*	*TX*	*CO*	*OH*	*TX*
Sustainable economic development	Support	Support	None observed	Support	None observed	None observed
Turnout	None observed	Support	None observed	Support	None observed	None observed
Socio-economic status	None observed	None observed	None observed	None observed	None observed	None observed
Percentage of owner-occupied homes	None observed	None observed	None observed	None observed	Support	None observed
Household value	None observed	None observed	Support	None observed	None observed	Support

Source: Author generated

Sustainable Economic Development and Local Defiance

Data in Ohio and Colorado show a connection between a city's support for sustainable economic development and local defiance towards oil and gas development. The 'green' motivation seemingly motivated skepticism in Longmont and Fort Collins, Colorado, and skepticism about Loveland's agreement with Anadarko. It also motivated elected leaders in Munroe Falls and Broadview Heights, Ohio, to enact local oil and gas policies that impeded the state's extractive goals. This finding is not groundbreaking and is supported in the applicable literature (Davis 2012). It suggests that the concerns about environmental protection are enough for some communities, especially when ballot initiatives can be used to set city policy, to take on the political and financial risks associated with challenging the state's preemptive authority (Fisk 2016).

Anecdotal evidence from other local leaders seemingly affirms this association. Oberlin, Ohio resident Sam Rubin, who led an effort to ban fracking within the city, explained that he was motivated by a desire to protect the environment (Rubin as quoted by Miller 2013). Anthony Capretta, a council member for the city of Brunswick, explained that he voted in favor of an anti-fracking resolution because of environmental concerns (Capretta as quoted in Lisik 2013). A second Brunswick City Councilman (at-large)

John Spon, the law director for the city of Mansfield, Ohio, defended his city's restrictions on fracking as necessary to protect its drinking water supply. He added that the ODNR regulatory framework is illogical and inadequate (Spon as quoted by Marshall 2012).

Local Defiance and Owner-Occupied Homes

Data suggest an association between challenging the state and the rate of owner-occupied homes. This association was only observed in Ohio. Part of this association is likely due to some of the political differences between homeowners and renters. DiPasquale and Glaeser (1999) observed that approximately 77 percent of homeowners reported that they had voted in a local election, compared with slightly more than 50 percent of renters. Additionally, they found that homeowners were more willing to engage in community-wide problem-solving efforts. Approximately 40 percent of homeowners reported that they had participated in addressing local problems, compared with a quarter of renters. Homeowners were also typically more willing to invest more of their time and energy into acquiring local amenities relative to renters. Each suggests that part of local defiance may be related to residents who stand to lose 'something' if a well pad or site is located near or in their community.

The link between home ownership and oppositional state-local relationships may also relate to differences in social capital within communities. DiPasquale and Glaeser write that home ownership impacts citizens' political behavior by positively influencing "the formation of social capital, and much of the influence of homeownership occurs because homeownership increases community tenure" (4, 1999). Putnam's (2000) classic work added that as a community's social capital 'score' increases, residents are more likely to become engaged with public affairs, trade and exchange information (such as risk evaluations) with other members of the community, and cooperate in terms of solving local problems. An industry-related earthquake, spill, or the potential of drilling in a park may activate these networks and prompt residents to quickly exchange information, cooperate with one another, and push their city to limit production (see also Hempel 2009; McKinney and Harmon 2002).

Local Defiance and Home Values

The relationship between defiance and home values makes intuitive sense as homeowners make important decisions about how development may impact their home's value (See Lubell 2002). Defiance, in this sense, may be part of a political strategy a city adopts to secure a competitive advantage against neighboring communities or protect its property tax base (see also Fisk 2016). In other words, they may have little concern about environmental impacts, but they may support higher setback distances (at least compared with surrounding communities), because officials believe that additional protections would keep residential property values high, attract additional economic investment, or improve community development.

A second way that the threat of diminishing home values may contribute to municipal defiance is by impacting the degree to which citizens are mobilizing, sharing information, and pressing leaders for policy change. Because nearby oil and gas production may negatively affect property values directly (based on well proximity) or indirectly through public safety, and quality of life, the benefits of local government pushback may become greater than its costs and therefore encourage engagement (Heinkel-Wolfe and Brown 2012).

Responses to drilling tied to housing characteristics may also relate to differences in wealth and affluence. Previous research by Hamilton (1995), for example, found that minority groups and poorer neighborhoods are less likely to overcome collective action problems (see also Ostrom 1990). A similar logic is observable relative to permitting and other siting issues. If permitting relative to well locations and site operations involves public input, then differences in community-level organization and difference in public participation levels may also lead to locational decisions that favor neighborhoods with lower rates of home ownership or lesser home values (see also Hamilton and Viscusi 1999; Opp 2012).

Flower Mound, Texas, illustrates the potential association between municipal defiance and drilling policies. The average Flower Mound home is valued at more than $250,000. However, when homes are situated within 1,000 feet of a well, the city estimated that their value decreased by between 3 and 14 percent. The analysis also noted that visual buffers, setbacks, and barriers moderated the financial impact of nearby wells; however, they did not eliminate it. Armed with this data, the city of Flower Mound passed a 1,500-foot setback in 2011, a de facto ban, according to industry (Integra Realty Resources 2010).

Recent empirical studies have also found a complicated relationship between well proximity and home values. A sampling of recent studies is noted, as follows:

- In Washington County, Pennsylvania, Gopalakrishnan and Klaiber (2014) observed a relationship between proximity to wells and a small drop in property values across the board.
- In Pennsylvania, Muehlenbachs, Spiller, and Timmins (2015) documented a negative relationship between homes that were dependent on groundwater and nearby shale development. The authors explained that "oft-debated risk to groundwater contamination has indeed materialized into a real impact" (4) The study also showed that homes dependent on piped water receive small benefits as a result of royalty and lease payments. Similarly proximate homes that have access to

publicly supplied piped water, on the other hand, appear to receive small benefits from that development. However, wells drilled recently do not contribute to this overall benefit.

(a similar version of this project is found in Muehlenbachs, Spiller, and Timmins 2013)

Note

1 Rosenstone and Hansen (1993) suggest that socio-economic factors such as home value can help explain turnout.

9 Summary and Recommendations

Summary of Comparative Findings

Domestic oil and natural gas production has undergone a renaissance in recent years, thanks in large part to horizontal drilling and hydraulic fracturing. Passionate political debate at all levels of government has followed. Proponents are quick to cite the prospect of good-paying jobs, growth in support industries, economic development, lower energy costs, and national security. Data in Chapters 4 through 6 suggest that supporters within Colorado, Ohio, and Texas explain their support in similar terms. But for others, including those described in the preceding chapters, domestic oil and gas production engenders powerful concerns over how it impacts drinking water supplies, public safety, air and water quality, quality of life, a neighborhood's character, home values, and public health. State and local skeptics argue that these concerns outweigh the economic benefits and are enough to justify additional oversight and regulation (Davis 2012; Fisk 2016).

The institutional and regulatory landscape for operators in 2017 is decidedly less laissez-faire compared with the 'boom's' early days. Despite different locations, politics, and historical experiences with oil and gas, the case studies demonstrated a number of similarities among the states and offer some generalizability to other jurisdictions participating in the oil and gas debate. In Colorado, Ohio, and Texas, lawmakers have passed an assortment of policies aimed at encouraging development, protecting the environment, and balancing state and local concerns. They have also tried less traditional mechanisms to reduce intergovernmental conflict, including legislative study groups, task-forces, and ballot amendments. In each of the states, local governments were often willing partners to state leaders and industry, and worked within applicable regulatory frameworks. This finding makes intuitive sense, because many cities and counties use their limited resources to compete for economic investment and growth opportunities, were aware of the risks of defiance, and believed that oil and gas development yielded positive economic outcomes (Fisk 2016; Kantor and David 1988). A Weld County, Colorado, official described his county's relationship with the state as mostly collaborative—and that local officials are aware of the challenges associated with managing oil and gas development. The official observed, "Oil and gas development is a hot topic and can be a political football for some. Trying to balance the interests of the oil and gas development with the interests of the public can be quite difficult. The Colorado Oil and Gas Conservation Commission (COGCC), for example, has a dual mandate. On one hand, they have to protect the interests of the mineral owner who by law has a right to extract their minerals but, on the other hand, they have to protect the interests of the public."

The case studies revealed a number of additional commonalities among the three states. In each of the case studies, oppositional intergovernmental relationships were much more likely to command media attention and prompt reactions from state and industry officials. In Colorado and Ohio, the respective State Supreme Court ultimately resolved the question of local control, whereas in Texas, legislators passed HB 40. Finally, in each of the states, stakeholders have actively debated oil and gas. Through these debates, claims of economic revitalization, job growth, and new revenues have been met with counterclaims that oil and gas development threatens to degrade the environment, is a menacing danger to public safety and health, and has ruined the charms of many communities. In each of the states, local defiance was likely the amalgam of idiosyncratic local characteristics and preferences, costs and benefits, state policies, and the location of development.

Despite greater regulatory controls at the state level, intergovernmental tensions have continued for a variety of reasons. Some strain is likely the result of the geology and location of oil and gas deposits. Because extraction can only take place near the deposits, the costs and benefits development are often concentrated in some areas and not in others. For communities in Colorado, local pushback was associated with a stronger municipal commitment to sustainable economic development and to voter turnout. For Ohio, higher home ownership rates were linked to local governments' willingness to challenge the state's preemptive authority. In response, leaders in both states successfully litigated against municipal defiance, although activists have suggested that the fight is far from over. Defiance, in Texas, appeared related to real and perceived threats to home values and as a result of localized focusing events. Texas lawmakers responded through the passage of HB 40, which has, thus far, prevented the involvement of state courts. The new law narrowed the authority of Texas's cities to regulate oil and gas development, outlawed municipal bans, and established a commercially reasonable standard for local regulation.

State Summaries

In Colorado, the state's Oil and Gas Conservation Commission oversees oil and gas development. Despite having a limited number of options available to them, several communities have injected themselves into the fracking fracas. Cities including Fort Collins, Boulder, Longmont, and Broomfield have enacted bans, issued temporary suspensions on drilling, or otherwise impacted where drilling may occur within their city limits (usually through zoning policies). In these communities, resistance appears linked to concerns about development's environmental and public safety. Other communities in Colorado have adopted more 'middle of the road' actions similar to those taken by the city of Loveland. Finally, the majority of Colorado cities, such as Greeley, Centennial, Aurora, and Windsor, have adopted a cooperative tone with COGCC regulators and industry (Udell 2013).

Ohio communities have also participated in intergovernmental conversations with state leaders about the scope and pace of development within cities and towns. Before the State Supreme Court's decision in *Munroe Falls*, local governments varied widely to the extent that they regulated oil and gas development and whether they restricted the locations available for drilling. However, much like Colorado, most Ohio substate governments have worked within the ODNR regulatory framework, which was clarified in the *Munroe Falls* decision. Ohio data points to home ownership as a key factor in the decision to challenge the state.

Like their counterparts in Ohio and Colorado, Texas cities are active actors on the intergovernmental unconventional oil and gas stage. Before HB 40, a quarter of Texas cities passed zoning policies to regulate oil and gas operations, often in an attempt to protect property values and quality of life. Even after HB 40, cities in Texas, assuming their regulations are commercially reasonable, may still enact zoning regulations and other oil and gas standards. Mansfield, Texas, for example, has banned oil and gas operations during weekends and city holidays. Several Texas communities have setback distances that remain at 1,500 feet, essentially a ban according to industry. Data suggest that oftentimes these stricter policies are associated with communities that have higher median home values and focusing events.

Table 9.1 displays the summarized results:

Table 9.1 Factors Associated with Intergovernmental Conflict, by State

	Colorado Defiance	*Ohio Defiance*	*Texas Defiance*
Sustainable economic development/green cities	Support	Some support	Some support
Voter turnout	Some support	Some support	No support observed
Per capita income	No support observed	No support observed	No support observed
Percentage of owner-occupied homes	No support observed	Some support	No support observed
Median household value (natural log)	Some support	No support observed	Support
Focusing events	Support	Support	Support
Lack of trust	Some support	Support	No support observed

Observations from Colorado lawmakers offer additional insights as to the causes of local defiance. Representative Joann Ginal, D-Fort Collins, noted that "fear is driving communities to bans and moratoriums, and fear shouldn't be the motivation" (Ginal as quoted by Jaffe 2014). Another state lawmaker in Colorado reflected that part of Fort Collins's reluctance to accept fracking relates to its brand: "For municipalities such as Fort Collins, one of [the] most important 'brands' is the quality of life that exists here and that [the] community is a healthy and safe place to live, work, and play, with open spaces, natural resources, and opportunities to be outside."

Yet, a representative of Weld County (home of Greeley) described oil and gas in a much more positive way, "I think that since technological advances have allowed the industry to find new plays and reach out to new areas where drilling hasn't typically happened the fear of the unknown kicks in. If hydraulic fracturing was really that bad, why does Weld County's population continue to increase? Why is our unemployment rate one of the lowest in the state? Property values continue to go up, not down. Surveys have shown that people that live in Weld like the quality of life."

Identifying Best Practices

Subnational governments are often the final deliverers of public services and have become entangled in the implementation of many other policies. The fracking fueled oil and gas renaissance is one such issue. It is multi-tiered and multi-dimensional but also idiosyncratic to particular contexts.

To effectively identify solutions, stakeholders must understand their co-dependencies, strengths, and the motivations behind local defiance. Local communities, for example, cannot adequately address environmental spillovers and externalities because ecological issues do not follow jurisdictional boundaries and can impact multiple communities in disparate ways. Conversely, states' policies that seek statewide benefits but fail to recognize the localized impacts of environmental harms are typically ineffective, prone to become inefficient, and often contribute to intergovernmental frustrations (Krane, Rigos, and Hill 2001; Ostrom 2009; Zimmerman 2012).

Effective intergovernmental management will rely on flexibility, organizational learning, and the deployment of nimble but strategic policy instruments. It must also comply with existing intergovernmental boundaries and the legal mechanisms that lead to legitimate policy change. In short, not all of the recommendations that follow will be transferrable across state lines. However, many of the recommendations do not require formal policy change but rather push stakeholders to invest in partnership and trust building. A strong, strategic, and more reflexive state-local partnership may not eliminate intergovernmental conflict, but it can lead to more effective, responsive, and inventive governance arrangements that assuage local fears, maximize benefits, and minimize costs.

Consider Improving Intergovernmental Relations

Industry has resolved many of the technical issues associated with fracking and production. Yet, a more complicated set of governance issues remain. Ranging from purely local concerns about the routes taken by heavy trucks and high levels of dust and noise to a general concern about urban oil and gas development, cities are increasingly injecting themselves into the fracking debate (Fisk 2016). If industry and state leaders better understand the sources of local concerns, then it is possible for them to begin the difficult task of identifying context-specific governing solutions.

The fracking debate involves serious public health stakes, the power of local democracy, and the economic livelihoods of millions of Americans. This complexity demands an all-hands-on-deck approach and the identification of new regulatory approaches that can leverage the strengths of federal, state, and local governments. A brief (and not comprehensive) synopsis of various strengths of each level of government is offered as follows:

- The federal government is needed to establish policies for issues that cut across state lines and to avoid an environmental race to the bottom.

(Konschnik and Boling 2014)

- States and industry often possess the expertise, resources, and legal authority to establish and enforce performance standards. States have also historically overseen natural resource development. State-level agencies, such as state oil and gas commissions, can focus statewide attention on a problem and create opportunities for local officials to utilize their discretion and judgment (Elmore 1980). Yet, because they are far removed from problem-solving pathways, it is difficult for higher-level officials to actually solve problems.
(See also Hupe 2011; Pressman and Wildavsky 1984)

- Local governments provide citizens with more access to the pathways of power, regulate land use and public safety, and may serve as repositories of local knowledge, which can inform more effective political and intergovernmental solutions, especially as they relate to the environment (Fiorino 2001, 2006; Klyza and Sousa 2007; Peterson 1995). Owens and Zimmerman (2013), for example, examined the relationship between organizational 'size' and efficacy of wetland protection efforts. Their data indicated that decentralization offered citizens additional opportunities to engage in the political process and shape wetland policy. Local engagement efforts, the authors note, also contributed to greater levels of public support for robust wetland protection.

Consider Capacity Building

Kettl (2005, 2002) argued that the foundation for effective governance begins with interactive relationships and from opportunities in which participants learn from one another (see Moe 1989). Organizational and policy learning may come from past experiences or from deliberate information exchanges in which stakeholders share accessible/relevant information, data, or experiences. Importantly, when state and local lawmakers communicate with the intent to improve their understanding of the challenges faced by their counterparts, better intergovernmental outcomes are possible (Agranoff and McGuire 2003). Research by Lovrich and Newman (2004), for example, found that when Washington state lawmakers had previous experience with or for local governments, they were typically more receptive to local interests. Hays's (1988) study of state-local transportation policy implementation in Iowa also demonstrated the importance of intentional capacity-building efforts. Hays (1988) attributed a positive state-local relationship (even after decisions against the local government) to the multiple opportunities during the planning and implementation process for local officials to communicate with state administrators. The agency also strategically cultivated a reputation for independent and professional decision-making.

A useful framework for categorizing state-local relationships is found in Scheberle's (2004) work, which identified four potential federal-state intergovernmental relationships. According to Scheberle (2004), intergovernmental relationships may be based on varying levels of trust and involvement with one another. Ideally, state-local relationships would be characterized by high trust and high involvement. By working towards high trust/high involvement relationships, stakeholders would have the capacity to search out new ways to solve multi-dimensional public challenges. These relationships also rely upon and seek out learning opportunities for participants to exchange information, receive technical assistance, offer and provide consultation, and receive guidance. Given that multi-dimensional factors such as breakdowns in trust and financial fairness are often the impetus driving defiance, high trust/involvement intergovernmental partnerships are imperative.

The intergovernmental system addressing oil and gas production is best described as fragmented. Building meaningful intergovernmental capacity related to managing oil and gas extraction therefore begins when stakeholders make a commitment to interacting with one another. Interactions, whether in person, phone, or online, should be devoted to improving stakeholders' understanding of the context-specific costs and benefits of oil and gas development, acknowledging one another's respective responsibilities and concerns, appreciating the causal mechanisms behind local defiance, and building the foundations necessary for more effective forms of collaboration. Collaborative relationships are further strengthened by the open and honest exchange of knowledge and context-specific information. This may be done through taskforces, workshops, conferences, hearings, and other meetings that meaningfully bring together state and local stakeholders. Importantly, if recommendations/policies are to be made, local input must be respected and recognized.

Whatever the design, the effectiveness of capacity-building efforts is often related to a number of principles that encourage meaningful collaboration; these principles are summarized in Table 9.2. These types of mutually supportive characteristics should be embedded into institutional processes, thereby facilitating additional opportunities for state-local trust-building and communication (Bardach 1977; Goggin et al. 1990; Sabatier 1986).

Without these important conversations, trust-building activities, and interactions among stakeholders, the causes of local defiance will remain unknown or misunderstood. As such, rather than designing effective and collaborative solutions, policy solutions will remain ineffective. A Weld County official observed that productive state-local-industry relationships involve "good planning, effective communication and creating an atmosphere and regulatory framework that provides protection to health, safety and welfare of the public but also allowing the free market to work so that companies can continue to invest in new technology and ingenuity."

Table 9.2 Collaborative Best Practices

Principle	*Principles in Practice*
Identifying shared goals	Is the goal to reduce oil and gas's environmental footprint? Is the goal to improve public safety and health? Is the goal to rebuild intergovernmental trust? Is there common ground for participants to identify effective and plausible solutions?
Real and perceived fairness	Do the standards apply to all groups equally, even those that have limited legal authority to engender change? Do stakeholders perceive fairness in permitting processes or other regulatory decisions? How will collaborative decisions be made (i.e., consensus, super-majority, or majority)? Is there a basic set of facts?
Scope	Do all participants agree upon the scope of issues to be discussed during the meeting(s) or other collaborative effort? Is the scope too large and broad for effective problem solving (i.e., to reduce all local concerns related to oil and gas development)? Is it too narrow (i.e., concern about one well or well pad)? Is it too vague and ambiguous for effective problem solving (i.e., goal is to improve local democracy)?
Expectations	Are performance and process expectations set? If those expectations are not agreed upon, then participants may hold dramatically different expectations about their role as part of the collaboration. For example, is the purpose to expand local regulatory flexibility or to share information? Are decisions advisory or binding?
Performance management	Intergovernmental capacity-building may also be used to identify and agree upon methodologies and goals to measure effectiveness, costs, and benefits. Performance could also be assessed during pre- and post-drilling tests. Other important questions to consider are how often measurements must take place, who may be given access to the results, and how the data is shared, especially with local first responders?

Source: Gratton and Erikson (2007)

Organizations that might serve as hosts or organizing bodies might include: state municipal leagues, state-city management/mayor/council member associations (such as the Texas City Managers Association or the Ohio City and County Management Association), state energy offices or regulatory bodies, or other ad-hoc organizations (such as the state-local task force utilized by Governor John Hickenlooper in Colorado), regional councils or bodies, and universities. In other cases, intergovernmental capacity-building may need to be done on a more personal level through deliberate trust-building efforts:

Recommendation 1: Consider increasing intergovernmental (state-local) deliberative capacity through meetings, focus groups, etc.

Consider Engaging Stakeholders, Especially at the Local Level

Another potential way to improve intergovernmental relations may be found in research dedicated to civic environmentalism. As an approach, it emphasizes the important role that decentralized institutions play in securing favorable environmental outcomes and building political efficacy. Increasing citizen deliberation in local and state lawmaking and/or regulatory bodies via the principles of civic environmentalism may also generate sizable social benefits such as greater levels of inter-organizational trust (John 1994; Leighninger 2006; Lemos and Agrawal 2006; Putnam 2000). Through deliberate negotiation, collaboration, and learning, civic environmentalism assumes that participants build the social capital needed to identify areas of agreement and overcome areas in which there is disagreement. An official from the city of Fort Worth added, "Bring industry into the conversation and let them represent themselves, let the people represent themselves—through the conversation they can disagree but the conversation can inform the formation and adoption of better public policy." A modified civic environmental model that focuses on the local aspects of oil and gas production could address the following:

- Identifying and agreeing upon a basic set of facts/knowledge related to oil and gas production, flowback, horizontal drilling (i.e., what is it, what are its context specific costs and benefits, and how is it similar/different to other extraction techniques).
- Identifying and agreeing upon a basic set of rules and values that may guide negotiations (such as Longmont, where even after a lawsuit the city and industry were able to work together).
- Identifying and better understanding oil and natural gas's policy dimensions in that particular jurisdiction (i.e., are citizens opposed to oil and gas development based on environmental concerns, dust/noise, the timing of operations, crime, blight, aesthetics, spills and other accidents, home values, etc.). If these 'dimensions' can be specifically identified and elucidated, targeted state or local responses or lobbying efforts become possible.
- Identifying and articulating what regulatory, contractual, and policy options are available to state and local leaders. Participants would need to understand their array of available policy instruments in the context of an intergovernmental system that protects private property rights, protects the environment and public safety, and wants to grow the economy and become energy independent.

Agreeing upon a basic set of facts and building social capital could yield a number of governing benefits. It may be, for example, that residents are not opposed to fracking or drilling but do not trust a particular operator or the opaqueness with which the operator handles emergencies (fire or spills). Similarly, deliberative citizen engagement may also show that citizens are not opposed to development but are against a well in a particular location or flaring at a specific time. In fact, a breakdown in trust between local leaders and an operator seemingly motivated activists in the cities of Arlington and Denton, Texas, and Longmont, Colorado, to push for more local control over development. If the cause of opposition is grounded in operator distrust, then new state policies, best practices, or management plans that strategically target information sharing or emergency preparedness might be devised rather than broad prohibitions on local land use authority.

Intra-organizational efforts may also look to how a jurisdiction evaluates the environmental and/or economic impacts of oil and gas:

- Identifying and agreeing upon a common set of goals and policy areas in which compromise with industry or the state is a possibility. Goals may range from restricting all oil and gas development to better information-sharing among stakeholders. Similarly, a deliberative exercise within a state regulatory body may reveal in which policy domains they believe compromise is possible.
- Efforts at working with citizens might be designed so that they better understand state-local legal frameworks and the likelihood of long-term success should they opt to support local actions that defy state policymakers and/or push the locality into adopting an ordinance that challenges the state's preemptive authority. This may have been a particularly useful option in cities such as Denton, Texas, in which citizens voted to ban fracking (but not all oil and gas production).
- Deliberative efforts might also be particularly fruitful if local governments decide to pursue voluntary agreements with industry, in which operators agree to standards more stringent than what the state requires. By enhancing their own capacity, the local government will be better positioned to negotiate with industry and advocate on behalf of citizens. Loveland, Colorado, for example, utilized private agreements with oil and gas operators to achieve some of its goals. Importantly, while some local leaders were unhappy with the outcome, in a subsequent ballot initiative, city voters rejected a proposed set of oil and gas restrictions. Longmont also achieved its goal of removing the Rider Well only after negotiating with TOP Operating.

Oil and gas impacts communities differently, even in states that have highly centralized permitting processes. One option is the use of a local liaison. In Weld County, Colorado, the position entails disseminating oil and gas information to residents. This individual is also expected to be a technical resource to interested stakeholders. The liaison may be well positioned to be an independent information conduit among state regulators, industry, and local authorities and able to assist bridging gaps and building intergovernmental problem-solving capacity.

Recommendation 2: Consider increasing democratic deliberation at the state and local levels.

Recommendation 3: Consider utilizing local liaisons for communicating residents' concerns to oil/gas operators, or for bolstering emergency management and planning functions.

Consider a Regional or Shale Play Approach

Shale plays and local governments are not created equally. Some cities, such as Youngstown, Ohio, and Fort Worth, Texas, welcome outside investment and the economic potential of unconventional oil and gas production. In others, citizens have pushed their municipalities into showdowns with state lawmakers. The question of local control, central to current intergovernmental debates, has compelling arguments on both sides. Those seeking more centralized forms of natural resource management suggest that one-size-fits-all state approaches lead to greater economic efficiency, lower transaction costs, and a more stable regulatory environment that encourages investment. Yet, by devolving authority and entrusting local or regional actors with greater discretion and flexibility, states may provide communities with the authority to more creatively and effectively govern oil and natural gas (Barzelay 2001, 1992; Eggers and O'Leary 1995; Gainsborough 2003).

One possible solution that capitalizes on the strengths of state and local approaches is to identify ways that regional actors or coalitions can work with industry and state regulators. Although this option capitalizes on local flexibility and economies of scale, it would also encounter a variety of challenges:

1. Adopting regional approaches may require state intervention before meaningful engagement and cooperation by local governments.
2. Transitioning to a regional approach, oil and gas management would dramatically alter the status quo in many states and would likely encounter significant resistance from a variety of actors, including the industry.
3. Local governments enjoy significant land use authority and may be unwilling to cede it to a regional authority.

If stakeholders were able to overcome the aforementioned obstacles, a regional approach would offer a number of advantages. First, it would permit industry to minimize efficiency losses that would result from differing systems of local regulation. It would, however, acknowledge local concerns, especially in differences between urban/suburban and more rural areas. Another way to improve the palatability of a regional approach is by carefully examining what aspects of oil and gas production lend themselves to regional solutions:

- Regional approaches may be particularly useful in identifying optimal transportation/traffic routes across metropolitan communities. This could be done to minimize traffic disruptions and maximize public safety.
- Regional air- and water-monitoring programs, which are already in use, may be utilized to monitor industry-related emissions (one particular challenge would likely relate to many of the exemptions listed in Chapter 2). The more heavily urbanized Barnett Shale region in Texas and Colorado's Front Range may be particularly interested in regional monitoring strategies, especially if they are close to being designated as a non-attainment area under the Clean Air Act.
- Emergency management functions may also be amenable to a regional approach. Regional coordination to better train and equip local firefighters to battle and mitigate blowouts, spills, and chemical fires may ease some of the fears undergirding local defiance. In the event of a spill or a well pad fire, these specialized resources could become available for those communities that entered into the inter-local agreement. Local governments in many states already have authority to engage in regional emergency management planning.
- Regional coordination may also be useful in addressing many of the social challenges during boom and bust times (Wiseman 2015). According to skeptics, for example, rapid oil and gas development has led to a number of undesirable social side effects such as increasing drug and alcohol abuse. Coordination by local governments and non-profits may be well positioned to address many of the 'social' disruptions associated with industrial development.

Recommendation 4: Consider regional approaches. Regional approaches could be used for emergency planning purposes and to ensure that regions are prepared with equipment and information (chemical lists) needed to address wellhead fires and other on-site emergencies.

Consider Voluntary Agreements

When crafting legislation, language matters. If language is overly restrictive, narrow, or otherwise limiting, it is likely that a state will create an expensive intergovernmental burden for local governments, industry, state-elected officials, and regulators. To respond to new conditions, cities will likely be forced to devote additional resources and time to lobbying state officials for the requisite local authority to act. Conversely, if the law is overly vague or fails to clearly identify issue boundaries, it is likely to precipitate litigation and/or conflict, because the courts will likely become the final arbiter of the law's definitional boundaries (Berman 2003; Matland 1995). In Texas, one local official noted that the language in HB 40 "makes it difficult to change ordinance and, right now, there is no benefit to continue to modify the ordinance as conditions, information, or data changes . . . even when something may change, every little thing can be challenged under HB 40 as failing to be commercially viable . . . even things that were not a big deal before the downturn."

One way that language may be clarified is through the greater use of voluntary agreements between local governments and industry or through memorandums of understanding (MOU) between stakeholders. Both Longmont and Loveland, Colorado, have worked with the state and industry to sign voluntary agreements that mitigated some of oil and gas development's impact to residents' quality of life externalities. Voluntary agreements typically involve local governments exchanging an expedited permitting process for land use concessions or the commitment by industry to adopt other best practices. Importantly, because they are entered into voluntary (and can be used to address one well or a set of wells), they can avoid legal challenges by industry and the state and also clarify vague and ambiguous language. Following are a number of areas that local governments may address via agreement:

- Pits: MOUs typically address an operator's pit practices within the jurisdiction, including freshwater or brine pits and setbacks from occupied structures, sensitive ecological areas, or areas of interest to the community.
- Technological advances: Under an MOU, locals may request that practices be revisited if new technological methods for construction or maintenance are developed and are available to the operator.
- Stormwater runoff: Under an MOU, locals may specify inspection powers (typically after a precipitation event and/or weekly) for berms and other features designed to mitigate stormwater runoff.
- Water supply and quality: In accordance with an MOU, a local jurisdiction may seek to include baseline water quality testing; this may include specific language addressing chemicals, distances, etc.
- Emergency planning, spill and release management: An MOU might allow a local government to stipulate that spills reported to the appropriate state agency also be reported to the local government. Local government may also request an emergency management plan from the operator.

- Private and public roads: Under an MOU, a local government may identify and charge back additional maintenance costs associated with industry's use of the right of way.
- Site aesthetics (lighting, paint, noise, weeds etc.): Under an MOU, a local government may specify and require specific site features.

For industry, agreements provide an additional layer of regulatory stability and the avoidance of a potentially damaging political fight with local leadership. Yet, private agreements do present some risks. First, they can involve time-consuming negotiations. Second, since they are confined to the terms of the contract, they do not produce a consistent citywide policy. In other words, the agreement may only apply to a specific well pad or a developer rather than all wells within the city.

> Recommendation 5: Consider encouraging the use of voluntary agreements such as memorandums of understanding.

Consider Designing or Permitting More Flexible Policy Instruments

Federal, state, and local policymakers each hold a unique set of political powers and tools. State involvement, for example, is needed for its expertise and ability to reduce transaction costs. Despite creating diseconomies of scale in terms of enforcement, local involvement is useful if it can improve the responsiveness of government and/or identify new regulatory approaches that respond to local needs. Additionally, communities often need the discretion and authority to effectively respond to localized impacts. Yet, to treat and understand oil and gas production as a singular topic (i.e., setbacks) is myopic. Effective governance involves the recognition that development has a set of interrelated issues (and fracking is just one of those issues). With this in mind, to effectively address the question of hydrocarbon governance, policymakers must consider a strategic mixture of policy and inducements administered by appropriate levels of government.

By varying the goals, structures, and processes relative to development's diverse set of geographical regions and issues, regulators may be able to better accommodate differences in local context within state regulatory programs. Land use and proximity-to-well concerns in highly populated areas along Colorado's Front Range, for example, could be given special consideration within a statewide regulatory program. Such programs could be modeled after the Clean Air Act, which has given non-attainment areas more time than those areas in attainment zones to achieve compliance with national standards. This would stand in place of state regulation that can often be described as one-size-fits-all.

In other cases, states might turn to performance instruments to provide further discretion to operators, who are willing to work with local governments or which are willing to adopt industry-wide best practices:

- States might consider reducing fees, taxes, or bonding amounts or permitting times in exchange for the industry's willingness to cooperate with local communities on topics from testing to response planning.
- Rewards might be related to decreasing permitting lag times and/or reporting requirements to those operators who again are willing to agree to local preferences or enter into a voluntary agreement.
- Rather than mandating specific uniform control technologies, pollution standards can be performance-based.

Recommendation 6: Consider taking advantage of performance targets and/or regulatory flexibility to incentivize developers to work with local leaders to achieve broadly defined standards.

Consider Stoplight/Traffic Light Types of Policies, Including Triggers

Stoplight approaches may help states and local governments better manage the impacts related to development located within cities and towns. By encouraging traffic light approaches, states may be able to mitigate the somewhat rare instances of local defiance. States governments in Colorado and Ohio have already adopted a similar approach for monitoring and regulating induced seismicity. The approach depends on real-time monitoring and management of impacts related to oil and gas wastewater injections in a particular area or meeting certain conditions.

Again, relying on Colorado's and Ohio's experiences with induced seismicity traffic light approaches as a guide, 'traffic lights' as applied to oil and gas production could involve the following:

- The site-specific factors that contribute to local defiance (i.e., why some wells spark high levels of environmental concern, emergencies/focusing events, and impacts to home values and quality of life).
- How to estimate and measure impacts. This could relate to the overall number of wells within the community, the number of wells in a specific neighborhood, proximity to occupied structures, and the number of trucks, etc.
- What severance tax rate should be collected (i.e., when oil and gas prices are lower, industry's effective rate could correspondingly drop and when prices are higher, the rate could also increase).

Lawmakers considering a triggering approach should consider the following:

- What should be the triggers for the stoplight policies (i.e., what leads to a green, yellow, and red light)?
- How to mitigate the impacts of oil and gas development (i.e., what gets triggered). Possible responses could be limits on the volume of truck traffic, additional site security and aesthetics measures, the placement of wells near vulnerable populations such as schools, or reductions in the number of wells operating within a community, etc.

In short, if specific and agreed-upon thresholds are reached in any of the foregoing scenarios, some policy option or rate would be triggered. Such an approach may have aided efforts in Denton, Texas, to balance state and local goals. Here, the number of wells and their locations near homes seemingly contributed to efforts to ban fracking within the community.

> Recommendation 7: Consider the use of legislative/regulatory triggers (i.e., if conditions exceed X, then local governments may do Y, etc.). This may include the use of stoplight approaches based on oil and gas's impacts to communities.

Consider Capacity-Building Assistance

In general, many of the economic benefits of oil and gas development accrue to the state. Although many local governments do directly benefit from production, they also more acutely experience its costs. Fort Worth, Texas, a community with more than 2,000 wells, strategically spends its industry-related revenues so that the people/areas most impacted by development are benefitting. A Fort Worth city official reflected, "We were really clear where the money would go . . . the cash that we generated went to benefit the city . . . under a park . . . under a road . . . we really try to spend it to offset the impacts of development."

Bonding requirements, fees, revenues, and state aid should adequately cover the costs of development, and help local governments plan for some of the costs associated with the 'bust' cycle of oil and gas extraction and some of the social costs. Additional trainings may be needed so that local officials can properly track revenues, document activity and total environmental and social costs, manage records, and submit paperwork to the appropriate state agency.

> Recommendation 8: Consider developing impact/revenue-sharing models so that local governments are more adequately compensated for the short- and long-term costs of development.

Consider Incubation Times for Local Governments and Industry

Bedsworth and Hanak (2013) argued that local compliance with state requirements and other mandates are strengthened when they are accompanied by long incubating and research periods. During these 'periods,' local governments or industry use the additional time to augment their organizational expertise, develop new capacities, and search for new and innovative options in an effort to meet the new standards. In addition, incubation times may also help avoid future conflict, by giving stakeholders the time to find avenues for cooperation, participate in state rulemakings, and trust building (see Alm, Buschman, and Sjoquist 2011). Bedsworth and Hanak (2013) point to California's waste diversion standards for local governments as an example. The mandate followed years of state encouragement and prodding. However, because it was coupled with ample lead time, local governments had the opportunity to build the organizational competencies needed to implement the program in a way that took their context into account. The 'waiting period model' appears to have been somewhat successful, with more than 50 percent of communities meeting the state standard (Bruhl, Linder, and Sexton 2013).

Although policymakers in Fort Worth did not rely upon a formal incubation time, they did note that state law gave them the latitude and discretion to research the impacts of oil and gas development within the city. The city spent approximately $1 million to commission a panel of industry and independent experts. Through research, the city settled on its current setback distance. In fact, their testing showed that 600 feet was a safe setback limit. A Fort Worth official noted, "We also learned that some local gas stations were giving off more air particulates than were recorded at some well sites." (Personal Communication, 2014). This particular approach may be useful for communities seeking to identify ideal setback distances or other policies that mitigate environmental impacts (assuming they have legal authority to adjust their setback distance). For jurisdictions in which setback distances are reserved to state regulators, research may be used in procedural hearings and other rulemaking efforts or as leverage during lobbying efforts. Studies and other planning opportunities may also be targeted towards emergency responses and regional coordination. To be clear, the use of incubation periods for research and coordination should only be pursued if the local government has the organizational capacity to conduct meaningful and accurate research. If requisite capacity is not available, a number of options are still available:

- Local governments may consider working with nearby universities.
- Local governments could coordinate and work towards a regional research project or study.
- Local governments could work with a trusted third party.

- Local governments may become more involved in research being conducted by state officials.

Incubation periods may be used to study local concerns related to:

- Studies on the appropriate setback and reverse setback distances from schools, homes, and other occupied facilities so that local governments may protect public health and safety. Similar research may done on flow and gathering lines.
- Studies on the potential dangers to nearby and regional water (drinking and/or recreational); this may include pre- and post-well testing.
- Studies on localized and regional air quality impacts of oil and gas development may be especially important if the local government is in a Clean Air Act non-attainment region, such as the Dallas-Fort Worth metro area.
- Studies and the development of plans to more effectively address emergency response processes; this may be of particular importance if the jurisdiction has many wells near schools, homes, medical facilities, etc. Similarly, an early-warning system (such as a text-message alert) for leaks, spills, and other emergencies may be developed and implemented based on residential location.
- Local governments may investigate and enter into interlocal emergency response units, which are specifically trained for oil and gas emergencies.
- Studies designed to answer questions on local and regional economic impacts (i.e., what is the expected job growth in the short- and long-term, home values, etc., during both the boom and bust portions of natural resource development).

> Recommendation 9: Consider the adoption of generous incubation times for local studies and capacity building when setting new state standards. However, establish sunset provisions for the studies so that research has set endpoints.

Consider the Implementation of Voluntary or Self-Regulation

Data highlight how spills and other accidents often precipitated acts of local defiance. Accidents are an endemic risk for any industrial activity, including oil and gas production. However, because unconventional oil and natural gas development is now taking place around millions of people, industry should take special precautions to minimize those risks.

One way that industry may become a better neighbor is through the continued development of industry-wide best practices, certification programs, and other forms of self-regulation (Abel, Stephan, and Kraft 2007). Industry, for example, may consider adopting voluntary initiatives by which operators collaborate with their peers and government to continuously improve their business operations and minimize their health, safety, and environmental disruptions for impacted residents. The chemical industry's Responsible Care program may serve as a template. The program includes guiding principles and several management programs designed to respond to community concerns, enhance product stewardship, and improve environmental health and safety (International Council of Chemical Associations 2016).

The EPA also has a voluntary program that may serve as a guide for state policymakers seeking to improve localized oil and gas production. Since 1993, the Natural Gas STAR program has encouraged partners to implement best management practices (BMPs) to reduce methane emissions. The program's BMPs address four core areas (EPA 2011a):

1. Providing information-sharing opportunities through workshops, fact sheets, newsletters, case studies, reports, and studies
2. Providing technical guidance through partner updates, technical reports, and other guidance from the EPA
3. Offering peer networking through webinars and other web-based communications
4. Recordkeeping services relative to emission reductions

Industry might also consider participating in labeling and certification programs. Labeling and third-party certification programs are also voluntary agreements that require producers to meet environmental or quality-of-life standards related to production or other activities. Standards should be the result of collaborative work of third-party actors, industry, and applicable units of government and administered/enforced through an independent organization.

A certification or self-regulation program that rewards industry for taking into consideration local concerns may be a particularly attractive option for communities in which multiple operators are operating on a large number of wells, such as Fort Worth, Texas. Here, rather than creating a voluminous amount of paperwork through individual voluntary agreements, a third party (state, local, or another group) may certify that the operators have met specific standards that minimize disruptions to nearby residents. One such example is the Center for Responsible Shale Development (CRSD). The center brings together energy producers and environmental groups and seeks ways to minimize environmental harms while allowing for the responsible development of natural resources. The center also lists 15 environmental performance standards and offers a certification program (Sustainable Shale 2016). These include flaring limitations, reductions of engine emissions, controls on storage tanks, and groundwater monitoring.

Recommendation 10: Consider encouraging participation in a program of voluntary self-regulation, such as the Center for Responsible Shale Development.

Consider the Use of More Rigorous Disclosure Programs

Disclosure requirements vary significantly across the states (Fisk 2013; Heikkila et al. 2014; Jackson et al. 2011; Kraft, Stephan, and Abel 2011). As Fisk (2013) suggests, some states require operators to share trade-secreted chemicals, volumes, toxicity, and other baseline information with the public and regulators. Other states have considerably less rigorous information-sharing programs. However, the value of that data without context or a clear presentation is muted. Butcher (1998) observes that when a recipient is bombarded with information, it becomes difficult for him or her to sift through it and identify priority items. Additionally, Bae, Wilcoxen, and Popp (2010) suggest that how the information is presented or shared with interested parties contributes to its utility. When agencies or industry present data in a vague or overly technical way, or in an ambiguous manner, it loses much of its usefulness (Stephan 2002). Finally, Desvousges, Smith, and Rink's (1992) analysis of radon-testing programs pinpoints two conditions as necessary for valuable disclosure programs. The first relates to the availability of timely information. Second, information must be clearly stated, broadly disseminated, and salient in the minds of the intended audience. For fracking fluids, this may require new measures that take into account toxicity, volume, and frequency of fluids used, and the presence or absence of occupied structures and vulnerable populations. Once calculated, data should be presented in an easy-to-understand way (i.e., toxicity grades, etc.).

Disclosure programs need to be strategically designed and able to balance the demands of environmental groups with the needs of regulated firms (Stephan 2002). Firm hesitancy or resistance may be related by several factors that should be considered as regulators design specific disclosure requirements. First, operators may have serious concerns that regulators, activists, or other groups will distribute data to shame/punish industry, mobilize and activate opposition to industry, gain a political advantage in a variety of institutions, or justify the need for additional layers of regulatory oversight. Second, oil and natural gas operators may legitimately worry that disclosure will cost them their economic competitiveness and other advantages. This concern is particularly acute given the low prices of oil and gas. Finally, disclosure programs may leave environmentalists and other skeptics unsatisfied and lead to charges that the public is being greenwashed (Jackson et al. 2011; Mazmanian and Kraft 2009).

> Recommendation 11: Consider developing more rigorous disclosure programs that present information in non-punitive ways that is accessible and easy to understand.

Consider Incorporating Risk-Based Policies

Many aspects of oil and gas development can heighten risk perceptions among members of the public. Thus, understanding how residents are likely to perceive these aspects is an important first step for stakeholders, especially when development takes place in communities that are likely unfamiliar with production. The psychometric risk evaluation framework identifies six factors that raise an individual's risk evaluation (Slovic, Fischhoff, and Lichtensein 1980). Several are rephrased as follows:

1. Is the community or neighborhood familiar with the hazard?
2. Has the community voluntarily assumed the risk?
3. How controllable are the risks?
4. What is the real and perceived potential damage (i.e., magnitude)?
5. How culpable are humans?
6. Are vulnerable populations at risk?

By asking and then honestly answering several of these questions, it is possible that several instances of local defiance may have been avoided. In Longmont, for example, leaking fluids near a school prompted local defiance. Here, vulnerable populations (school-aged children) were at risk. Those motivated by this spill may have also perceived that their community did not assume the risk and were unfamiliar with them. The leak may have persuaded some residents that risks associated with development were too high. It should be noted that the Rider Well predated the school and many of the residents who voiced concerns to city council.

With 'risk' in mind, what may be additional areas that could prevent future local defiance?

1. Consider distinguishing between traditional setbacks and reverse setbacks. Reverse setbacks capitalize on the idea that citizens who occupy new structures located near existing wells are much more likely to have voluntarily assumed the risk. Local governments may also consider shorter reverse setbacks and larger traditional setbacks may preclude citizen unease about development.
2. Operators should take extra precautions if they construct wells or facilities located near vulnerable populations such as schools or hospitals.
3. Stakeholders, including the oil and gas industry and state regulators, should engage with neighborhoods so that residents have a greater familiarity with oil and gas operations. This may mean additional citizen meetings and presentations that go beyond the state-required public hearings.
4. Take steps to incorporate risk mitigation technologies to improve the steel casing gathering lines and the cement used in wells. The risks of drinking water contamination and methane migration are high when the steel casing used in gathering and flow lines leaks (can occur at the connections or through corrosion caused by caustic chemicals). The cement used during operations may also fail over time or when it shrinks or cracks, raising the likelihood of contamination or migration (Jackson 2016).

Recommendation 12: Incorporate strategic risk mitigation (as it relates to governance) thinking into decision-making.

Table 9.3 offers a summation.

Table 9.3 Summary of Recommendations

Recommendations	
Recommendation 1	Consider increasing intergovernmental (state-local) deliberative capacity through meetings, focus groups, etc.
Recommendation 2	Consider increasing democratic deliberation at the state and local levels.
Recommendation 3	Consider utilizing local liaisons for communicating residents' concerns to oil/gas operators, or for bolstering emergency management and planning functions.
Recommendation 4	Consider regional approaches. Regional approaches could be used for emergency planning purposes and to ensure that regions are prepared with equipment and information (chemical lists) needed to address wellhead fires, and other on-site emergencies.
Recommendation 5	Consider encouraging the use of voluntary agreements such as memorandums of understanding.
Recommendation 6	Consider taking advantage of performance targets and/or regulatory flexibility to incentivize developers to work with local leaders to achieve broadly defined standards.
Recommendation 7	Consider the use of legislative/regulatory triggers (i.e., if conditions exceed X, then local governments may do Y, etc.). This may include the use of stoplight approaches based on oil and gas's impacts to communities.
Recommendation 8	Consider developing impact/revenue-sharing models so that local governments are more adequately compensated for the short- and long-term costs of development.
Recommendation 9	Consider the adoption of generous incubation times for local studies and capacity building when setting new state standards. However, establish sunset provisions for the studies so that research has set endpoints.
Recommendation 10	Consider encouraging participation in a program of voluntary self-regulation, such as the Center for Responsible Shale Development.
Recommendation 11	Consider developing more rigorous disclosure programs that present information in non-punitive ways that is accessible and easy to understand.
Recommendation 12	Consider incorporating strategic risk mitigation (as it relates to governance) thinking into decision-making.

Source: Author generated

Conclusions

Who governs is one of the most venerable questions asked by public administration and political science scholars. Scholars, tracing back to Dahl (1961) and more recent federalism writers such as Bowman and Kearney (2010, 2011), characterize subnational political power as ephemeral, evolving, and shared. The Ohio, Colorado, and Texas case studies suggest that questions of power are at the center of local resistance to the shale boom and that it is shaped by state policy, shale characteristics, and the distribution of costs and benefits. Local preferences related to environmental protection, quality of life, emergency management, and economic factors also impacted the willingness to challenge the state's preemptive authority (Birkland 1997). These findings affirm Anton's (1989) conclusion that since political power must be shared, tensions between actors are inevitable (Bowman and Kearney 2012, 2011, 2010; Busche 2010; Krane, Rigos, and Hill 2001).

The answers to the fracking debate are likely to leverage the strengths of federal, state, and local governments (Ostrom 2010, 2009). When stakeholders acknowledge their interdependencies and one another's expertise, they create, design, and implement more responsive, strategic, and targeted public policies (Kooiman 1993; Feldman and Khademian 2002, 534).Throughout this project, readers can note several possible options that may calm local concerns and still permit industry innovation. Some recommendations may not be possible and would require legislative authorization. Others encourage stakeholders to focus on relationship building among the industry, state officials, and local governments. Such approaches would push organizations to learn and adapt once new information is acquired (Konschnik and Boling 2014).

Works Cited

Abel, Troy, Mark Stephan, and Michael E. Kraft. 2007. "Environmental Information Disclosure and Risk Reduction among the States." *State and Local Government Review* 39(3): 153–165.

Adams, William M. 2006. "The Future of Sustainability: Re-Thinking Environment and Development in the Twenty-First Century." http://cmsdata.iucn.org/downloads/iucn_future_of_sustanability.pdf

Agranoff, Robert, and Michael McGuire. 2003. *Collaborative Public Management: New Strategies for Local Governments.* Washington, DC: Georgetown University Press.

———. 2001. "American Federalism and the Search for Models of Management." *Public Administration Review* 61(6): 71–81.

Allen, Jason. 2014. "North Texans Protest Fracking, Earthquakes at Railroad Commission Meeting." *CBS*, January 21. http://dfw.cbslocal.com/2014/01/21/north-texans-protest-fracking-earthquakes-at-railroad-commission-meeting/.

———. 2013. "Mansfield Residents Upset about Fracking Near Their Homes." *CBS-DFW*, October 14. http://dfw.cbslocal.com/2013/10/14/mansfield-residents-upset-about-fracking-near-their-homes/.

Alm, James, Robert D. Buschman, and David L. Sjoquist. 2011. "Citizen 'Trust' as an Explanation of State Education Funding to Local School Districts." *Publius: The Journal of Federalism* 41(4): 636–661.

American Oil and Gas Historical Society. N.d. "Gasbuggy Tests Nuclear Fracking." http://aoghs.org/technology/project-gasbuggy/.

American Petroleum Institute. 2014. "Hydraulic Fracturing Unlocking America's Natural Gas Resources." www.api.org/~/media/Files/Policy/Exploration/HYDRAULIC_FRACTURING_PRIMER.ashx.

Anderson, Dave. 2014. "Paying to Pump: Missing Oil and Gas Severance Tax Revenue." *Boulder Weekly*, March 13. www.boulderweekly.com/article-12514-paying-to-pump.html.

Anderson, William. 1960. *Intergovernmental Relations in Review.* Minneapolis, MN: University of Minnesota Press.

Anton, Jamie. 2013. "City Joins Munroe Falls' Fight to Protect Home Rule: Sends State Message Cities Want More Control Regarding Urban Drilling." *The Royalton Post*, September 11. www.thepostnewspapers.com/north_royalton/local_news/article_4eb8d22a-a166-5e3a-ae18-1149b41c50b4.html

Anton, Thomas J. 1989. *American Federalism and Public Policy: How the System Works.* Philadelphia: Temple University Press.

Antonacci, Karen. 2016. "Does Colo. Supreme Court Ruling Mean Closing Longmont's Rider Well?" *Longmont Times-Call*, May 7. www.timescall.com/longmont-local-news/ci_29863005/does-colo-supreme-court-ruling-mean-closing-longmonts.

———. 2015. "Longmont Fracking Ban to Go before Colorado Supreme Court." *Longmont Times-Call*, September 21. www.timescall.com/energy-news/ci_28851304/colorado-supreme-court-accepts-longmont-fracking-ban-case.

Archbold, Carol A., Thorvald Dahle, and Rachel Jordan. 2014. "Policing 'The Patch': Police Response to Rapid Population Growth in Oil Boomtowns in Western North Dakota." *Police Quarterly* 17(4): 386–413.

Arenschield, Laura. 2014. "Fracking Industry Suing Over Drilling Bans." *The Columbus Dispatch*, November 21. www.dispatch.com/content/stories/local/2014/11/21/industry-suing-over-drilling-bans.html.

———. 2014a. "Ohio Fracking Water Reuse Questioned." *The Columbus Dispatch*, July 13. www.dispatch.com/content/stories/local/2014/07/13/fracking-water-reuse-questioned.html.

Arlington Tomorrow Foundation. 2015. "What We Do: Past Grants: FY 2013 Grant Recipients." www.arlington-tx.gov/arlingtonfoundation/what-we-do/past-grants/fy-2013-grant-recipients/.

Armendariz, Al. 2009. "Emissions from Natural Gas Production in the Barnett Shale Area and Opportunities for Cost-Effective Improvements." www.fe.doe.gov/programs/gasregulation/authorizations/2011_applications/exhibits_11-161-LNG/13_Barnett_Shale_Report.pdf.

Associated Press. 2015. "Judge Overturns Mora County's Drilling Ordinance." *Albuquerque Journal*, January 20. www.abqjournal.com/528970/news/judge-overturns-new-mexico-countys-drilling-ordinance.html.

Bachrach, Peter, and Morton Baratz. 1962. "The Two Faces of Power." *American Political Science Review* 56(4): 947–952.

Bae, Hyunhoe, Peter Wilcoxen, and David Popp. 2010. "Information Disclosure Policy: Do State Data Processing Efforts Help More Than the Information Disclosure Itself?" *Journal of Policy Analysis and Management* 29(1): 163–182.

Baker, Max. 2015. "Denton City Council Repeals Fracking Ban." *Dallas Star-Telegram*, June 17. www.star-telegram.com/news/business/barnett-shale/article24809482.html#storylink=cpy.

———. 2015a. "Denton Fracking Bill Sails through Texas Senate." *Dallas Star-Telegram*, May 4. www.startelegram.com/news/business/barnettshale/article20199849.html#storylink=cpy.

———. 2015b. "Governor Abbott Signs Denton Fracking Bill." *Dallas Star-Telegram*, May 18. www.startelegram.com/news/business/barnettshale/article21282813.html#storylink=cpy.

Balmert, Jessie. 2015. "Ohio Communities Want Control over Fracking." *Cincinnati Enquirer*, July 8. www.cincinnati.com/story/news/2015/07/08/ohio-communities-want-control-fracking/29876557/.

Banerjee, Neela. 2014. "Fracking Workers Exposed to Dangerous Amounts of Benzene, Study Says." *Los Angeles Times*, September 11. www.latimes.com/science/sciencenow/la-sci-sn-fracking-benzene-worker-health-20140910-story.html.

Bardach, Eugene. 1977. *The Implementation Game*. Cambridge, MA: MIT Press.

Barnes, Matthew. 2013. "Hydrofracking and Policymaking in New York Municipalities." Presented at the Annual Meeting of the American Political Science Association, Chicago, IL, August 2013.

Barnett, Marissa. 2014. "Denton's Ban Won't Stop Fracking Permits, Railroad Commission Chairwoman Says." *Dallas Morning News*, November 7. www.dallasnews.com/news/state/headlines/20141107-dentons-ban-wont-stop-fracking-permits-railroad-commission-chairwoman-says.ece.

Bartik, Alexander Wickman, Janet Currie, Michael Greenstone, and Christopher R. Knittel. 2016. "The Local Economic and Welfare Consequences of Hydraulic Fracturing." December 22. https://ssrn.com/abstract=2692197.

Barzelay, Michael. 2001. *The New Public Management: Improving Research and Policy Dialogue*. 3rd ed. Berkeley, CA: University of California Press.

———. 1992. *Breaking Through Bureaucracy*. Berkeley, CA: University of California Press.

Batheja, Aman. 2015. "Hegar: Stalling Oil Boom Cutting into State Revenues." *The Texas Tribune*, October 13. www.texastribune.org/2015/10/13/hegar-stalling-oil-boom-means-tax-revenue-coming-l/.

Baumgartner Frank R., and Bryan D. Jones. 1991. *Agendas and Instability in American Politics*. Chicago, IL: University of Chicago Press.

Bureau of Economic Analysis [BEA]. 2016. "Industry Data." http://bea.gov/iTable/iTable.cfm?ReqID=51&step=1#reqid=51&step=51&isuri=1&5114=a&5102=15.

Bedsworth, Louise W., and Ellen Hanak. 2013. "Climate Policy at the Local Level: Insights from California." *Global Environmental Change* 23(3): 664–677.

Begos, Kevin. 2014. "4 States Confirm Water Pollution from Drilling." *USA Today*, January 5. www.usatoday.com/story/money/business/2014/01/05/some-states-confirm-water-pollution-from-drilling/4328859/.

Bekiemphis, Victoria. 2015. "Florida City Bans Fracking." *Newsweek*, July 15. www.newsweek.com/florida-city-bans-fracking-do-not-run-embargo-embargo-embargo-untilsomething-353047. During the summer of 2015, the city of Bonita Springs joined the city of Coconut Creek in restricting the locations in which hydraulic fracturing may occur.

Berman, David R. 2003. *Local Government and the States: Autonomy, Politics, and Policy*. Armonk, NY: M.E. Sharpe Inc.

Berry, Jeffrey M. 1989. *Subgovernments, Issue Networks, and Political Conflict*. Boulder, CO: Westview Press.

Berwyn, Bob. 2013. "U.S. Rep. Jared Polis Co-Sponsors Bills to Regulate Fracking." *Summit County Voice*, March 16. http://summitcountyvoice.com/2013/03/16/u-s-rep-jared-polis-co-sponsors-bills-to-regulate-fracking.

Betsill, Michele M. 2001. "Mitigating Climate Change in U.S. Cities: Opportunities and Obstacles." *Local Environment* 6(November): 393–406.

———, and Barry G. Rabe. 2009. "Climate Change and Multilevel Governance: The Evolving State and Local Roles." In *Toward Sustainable Communities*. eds. Daniel A. Mazmanian and Michael E. Kraft. Cambridge: MIT Press.

Birkland, Thomas A. 2011. *An Introduction to the Policy Process: Theories, Concepts, and Models of Public Policy Making*. Armonk, NY: M.E. Sharpe Inc.

———. 1997. *After Disaster: Agenda Setting, Public Policy, and Focusing Events*. Washington, DC: Georgetown University Press.

Block, Melissa. 2014. "Drilling Frenzy Fuels Sudden Growth in Small Texas Town." National Public Radio, April 10. www.npr.org/2014/04/10/295332292/drilling-frenzy-fuels-sudden-growth-in-small-texas-town.

Boersma, Tim, and Corey Johnson. 2012. "The Shale Gas Revolution: US and EU Policy and Research Agendas." *Review of Policy Research* 29(4): 570–576.

Bohatch, Emily. 2015. "Ohio Supreme Court: Local Governments Can't Ban Fracking." *Athens Post*, February 17. www.thepostathens.com/news/ohio-supreme-court-local-governments-can-t-ban-fracking/article_78a5792c-b6cc-11e4-bd0b-7f239ccee436.html.

Bosso, Christopher. 2005. *Environment, Inc.: From Grassroots to Beltway*. Lawrence, KS: University Press of Kansas.

Boudet, Hilary, Christopher Clarke, Dylan Bugden, Edward Maibach, Connie Roser-Renouf, and Anthony Leiserowitz. 2014. "Fracking Controversy and Communication: Using National Survey Data to Understand Public Perceptions of Hydraulic Fracturing." *Energy Policy* 65(February): 57–67.

Bowman, Ann, and Richard Kearney. 2012. "Are U.S. Cities Losing Power and Authority? Perceptions of Local Government Actors." *Urban Affairs Review* 48(4): 528–546.

———. 2011. "Second-Order Devolution: Data and Doubt." *Publius: The Journal of Federalism* 41(4): 563–585.

———. 2010. "Subsidiarity in State-Local Relations." *Available at SSRN 1665467*.

Brady, William J., and James P. Crannell. 2012. "Hydraulic Fracturing Regulation in the United States: The Laissez-Faire Approach of the Federal Government and Varying State Regulations." *Vermont Journal Environmental Law* 14: 39.

Branstetter, Ziva. 2015. "Under Pressure? Do E-Mails Tell of Earthquake Information Sharing or State, Industry Interference?" *Tulsa World*, March 4. www.tulsaworld.com/earthquakes/under-pressure-do-emails-tell-of-earthquake-information-sharing-or/article_a7df8a5c-47f7–540c-af46-faa04765bc5c.html.

Brasier, Kathryn J., Diane K. McLaughlin, Danielle Rhubart, Richard C. Stedman, Matthew R. Filteau, and Jeffrey Jacquet. 2013. "Risk Perceptions of Natural Gas Development in the Marcellus Shale." *Environmental Practice* 15(2): 108–122.

Brookings Institution. 2011. "Sizing the Clean Economy: A National and Regional Green Jobs Assessment." www.brookings.edu/~/media/research/files/reports/2011/7/13%20clean%20economy/0713_clean_economy.pdf.

Brown, Katie. 2015. "States Well Ahead of EPA on Underground Wastewater Disposal Regulations." *Energy in Depth*.

Brown, Stephen P. A., and Mine K. Yücel. 2013. "The Shale Gas and Tight Oil Boom: US States' Economic Gains and Vulnerabilities." *Council on Foreign Relations*. www.cfr.org/united-states/shale-gas-tight-oil-boom-us-states-economic-gains-vulnerabilities/p31568.

Bruhl, Rebecca J., Stephen H. Linder, and Ken Sexton. 2013. "Case Study of Municipal Air Pollution Policies: Houston's Air Toxic Control Strategy under the White Administration, 2004–2009." *Environmental Science & Technology* 47(9): 4022–4028.

Brumfield, Sara. 2013. "Athens, Amesville Show Support for 'Home Rule' in Munroe Falls Fracking Appeal." *Athens Messenger*, September 24. www.athensohiotoday.com/news/athens-amesville-show-support-for-home-rule-in-munroe-falls/article_0e8f07ed-5c85–56d0-b422-d03b77dd4298.html.
Buchele, Mose. 2014. "Denton Voted to Ban Fracking: So Now What?" *National Public Radio*, November 6. https://stateimpact.npr.org/texas/2014/11/06/denton-voted-to-ban-fracking-so-now-what/.
Bunch, A. G., C. S. Perry, L. Abraham, D. S. Wikoff, J. A. Tachovsky, J. G. Hixon, J. D. Urban, M. A. Harris, and L. C. Haws. 2014. "Evaluation of Impact of Shale Gas Operations in the Barnett Shale Region on Volatile Organic Compounds in Air and Potential Human Health Risks." *Science of the Total Environment*, 468(January): 832–842.
Bureau of Economic Analysis [BEA]. 2016. "Industry Data." http://bea.gov/iTable/iTable.cfm?ReqID=51&step=1#reqid=51&step=51&isuri=1&5114=a&5102=15.
———. N.d. "Regional Data." http://bea.gov/itable/iTable.cfm?ReqID=70&step=1#reqid=70&step=30&isuri=1&7022=11&7023=7&7024=naics&7033=-1&7025=4&7026=08000&7027=-1&7001=711&7028=-1&7031=08000&7040=-1&7083=levels&7029=33&7090=70.
Bureau of Labor Statistics. 2016. "Employment and Output by Industry." www.bls.gov/emp/ep_table_207.htm.
———. 2016a. "Support Industries." http://beta.bls.gov/dataViewer/view/timeseries/CEU1021311201;jsessionid=03525FA776F6505EBA9E6551A47C3E79.tc_instance6.
———. 2014. "Unemployment." www.bls.gov/eag/eag.oh_youngstown_msa.htm (January 8, 2014).
———. 2013a. www.bls.gov/green/home.htm.
———. 2013b. www.bls.gov/bls/sequester_info.htm.
Bureau of Land Management. 2016. "Recent Oil and Gas Lease Sales." www.blm.gov/wo/st/en/prog/energy/oil_and_gas/recent_lease_sales.html.
Bureau of Ocean Energy Management. 2016. "Combined Leasing Report." www.boem.gov/Combined-Leasing-Report-January-2016/.
———. N.d. "Federal Offshore Lands." www.boem.gov/Federal-Offshore-Lands/.
Busche, Sarah. 2010. "Clean Energy Policy Analyses: Analysis of the Status and Impact of Clean Energy Policies at the Local Level." *National Renewable Energy Laboratory*. www.nrel.gov/docs/fy11osti/49720.pdf (August 2, 2013).
Bustillo, Miguel, and Daniel Gilbert. 2015. "Energy's New Legal Threat: Earthquake Suits." *Wall Street Journal*, March 30. www.wsj.com/articles/frackings-new-legal-threat-earthquake-suits-1427736148.
Butcher, Helen. 1998. "Information Overload in Management and Business." *IEE Colloquium Digest* 95(223): 1–2.
Cadwallader, Robert. 2015. "Mansfield Council Moves to Tighten Gas Well Ordinance." *Dallas Star-Telegram*, February 23. www.star-telegram.com/news/local/community/mansfield-news-mirror/mnm-news/article11053832.html.
Cama, Timothy. 2015. "FBI to Open Office in ND Oil Country." *The Hill*, March 5. http://thehill.com/policy/energy-environment/234793-fbi-to-open-office-in-nd-oil-country.
———. 2015a. "Judge Blocks Federal Fracking Rule." *The Hill*, June 23. http://thehill.com/policy/energy-environment/245933-judge-blocks-federal-fracking-rule.

———. 2015b. "Ohio Court Strikes Down Local Fracking Bans." *The Hill*, February 17. http://thehill.com/policy/energy-environment/232945-ohio-court-strikes-down-local-fracking-bans.

———. 2014. "Cheniere Signs 20-Year Deal to Export Gas to Spanish Company." *The Hill*, May 30. http://thehill.com/policy/energy-environment/207794-cheniere-signs-20-year-deal-to-export-gas-to-spanish-company.

Carr-Smyth, Julie. 2015. "Ohio Supreme Court Rejects Fracking Foes' Ballot Protest." *Houston Chronicle*, September 16. www.houstonchronicle.com/business/energy/article/Ohio-Supreme-Court-rejects-fracking-foes-ballot-6508770.php.

Cart, Julie. 2013. "New Mexico County First in Nation to Ban Fracking to Safeguard Water." *Los Angeles Times*, May 28. http://articles.latimes.com/2013/may/28/local/la-me-fracking-ban-20130529.

CBS Denver 4. 2013. "Hickenlooper Threatens to Sue Any Town, City That Bans Fracking." *CBS 4*, February 26. http://denver.cbslocal.com/2013/02/26/hickenlooper-threatens-to-sue-any-town-city-that-bans-fracking/.

———. 2012. "Colorado Springs OKs Drilling Despite Fracking Fears." *CBS 4*, November 27. http://denver.cbslocal.com/2012/11/27/colorado-springs-oks-drilling-despite-fracking-fears/.

Chavez, John. 2014. "Ohio's Natural Gas Boom brings Flurry of Pipeline Construction." *Toledo Blade*, December 7. www.toledoblade.com/business/2014/12/07/Ohio-s-natural-gas-boom-brings-flurry-of-pipeline-construction.html#kc1eVh42t0LlIqLU.99.

Ciesco, Tim. 2015. "Site of Arlington Fracking Fluid Leak Could Soon Reopen." *NBCDFW*, June 16. www.nbcdfw.com/news/local/Site-of-Arlington-Fracking-Fluid-Leak-Could-Soon-Reopen-307732141.html.

City of Akron. 2015a. "Executive—Mayor's Office." www.akronohio.gov/cms/site/172f2d8288f1a940/index.html.

———. 2015b. "Keep Akron Beautiful." www.keepakronbeautiful.org.

City of Arlington. 2016. "FY 2016 Proposed Budget." www.arlington-tx.gov/budget/wp-content/uploads/sites/19/2014/05/FY-2016-Proposed-Budget-Final.pdf.

———. 2016a. "Gas Well Drilling." www.arlington-tx.gov/cdp-gaswells/.

———. 2014. "Urban Drilling Presentations." www.arlington-tx.gov/cdp-gaswells/wp-content/uploads/sites/44/2014/12/Urban-Drilling-Presentations1.pdf.

———. 2011. "Gas Drilling Chapter." www.arlington-tx.gov/cdp-gaswells/wp-content/uploads/sites/44/2014/12/GasDrilling-Chapter.pdf.

———. N.d. "Staff Information." www.arlington-tx.gov/cdp-gaswells/staff/.

City of Broadway Heights. 2015. "City Council." www.broadview-heights.org/133/CIty-Council.

———. 2015a. "Mayor." www.broadview-heights.org/314/Mayor.

———. 2012. "Oil and Gas Well Information." www.broadview-heights.org/315/Gas-Oil-Well-Information.

City of Denton. 2016. "2015–2016 Annual Program and Services." www.cityofdenton.com/home/showdocument?id=26393.

———. 2016a. "FAQs and Agency Links." www.cityofdenton.com/departments-services/departments-g-p/gas-well-inspections/frequently-asked-questions.

———. 2016b. "Forms and Fees." www.cityofdenton.com/departments-services/departments-g-p/gas-well-inspections/forms-and-fees.

City of Fort Collins. 2016. "City Council." www.fcgov.com/council/.
———. 2016a. "Oil and Gas." www.fcgov.com/oilandgas/.
———. 2013. "Oil and Gas Informational Meeting." May 8. www.fcgov.com/oilandgas/pdf/presentation20130508.pdf.
———. N.d. "Fort Collins Oil and Gas Regulation Frequently Asked Questions." www.fcgov.com/oilandgas/faq.php.
City of Fort Worth. 2016. "City Government." http://fortworthtexas.gov/government/.
———. 2016a. "Gas Wells." http://fortworthtexas.gov/gaswells/.
———. 2011. "Fort Worth Natural Gas Air Quality Study Final Report." http://fortworthtexas.gov/uploadedFiles/Gas_Wells/AirQualityStudy_final.pdf.
City of Greeley. 2016. "City Council." http://greeleygov.com/government/city-council.
———. 2013. "Water and Fracking." http://greeleygov.com/Water/AnnouncementsDisplay.aspx?ID=200&Source (March 30, 2014).
———. 2011. "Oil & Gas Development Frequently Asked Questions (FAQs)." http://greeleygov.com/CommunityDevelopment/Documents/Planning/Forms%20and%20Handouts/Oil%20and%20Gas%20FAQ.pdf.
City of Longmont. 2016. "Oil and Gas Regulations in Longmont." www.longmontcolorado.gov/departments/departments-n-z/public-information/oil-and-gas-information/oil-and-gas-regulations-in-longmont.
———. 2016a. "Oil and Gas Well Scheduled to Close in August." www.longmontcolorado.gov/Home/Components/News/News/2899/661?backlist=%2Fdepartments%2Fcity-council.
———. 2013. "City Council." www.ci.longmont.co.us/city_council/index.htm.
———. 2013a. "Environment." www.ci.longmont.co.us/environment/index.htm.
———. 2013b. "Oil and Gas Information." www.longmontcolorado.gov/departments/departments-n-z/public-information/oil-and-gas-information/fracking-in-longmont.
City of Loveland. 2015. "Oil and Gas Development in Loveland." www.ci.loveland.co.us/index.aspx?page=2206.
———. 2013. "City Council." www.ci.loveland.co.us/index.aspx?page=67.
———. 2013a. "City of Loveland Community Sustainability Plan Version 2013-B." www.ci.loveland.co.us/modules/showdocument.aspx?documentid=11760.
City of Mansfield. 2016. "Mayor and Council." www.mansfieldtexas.gov/mayor-and-city-council.
———. 2015. "About Gas Wells." www.mansfieldtexas.gov/about-gas-wells.
———. 2015a. "Ordinance." www.mansfieldtexas.gov/sites/default/files/media/chapter114gaswelldrillingregulations.pdf.
———. N.d. "Gas Well Permitting Process." www.mansfieldtexas.gov/sites/default/files/media/gas_well_permitting_process.pdf.
City of Munroe Falls. 2016. "City Ordinances." www.munroefalls.com/index.php?option=com_content&task=view&id=116&Itemid=36.
———. 2015. "City Hall." www.munroefalls.com//index.php?option=com_content&task=category&sectionid=8&id=18&Itemid=35.
———. 2012. "Shale, Fracking, Oil and Gas Well FAQ." www.munroefalls.com/downloads/news/Fracking%20FAQ.pdf.
———. N.d. "City Council Passes Gas and Oil Resolution." www.munroefalls.com//index.php?option=com_content&task=view&id=227&Itemid=36.

City of Youngstown. 2016. "City Hall." www.youngstownohio.gov/city_hall/departments/council/council.aspx.

———. N.d. "Departments." www.youngstownohio.gov/city_hall/departments/litter/litter.aspx.

Cobb, Roger W., and Charles D. Elder. 1972. *Participation in American Politics: The Dynamics of Agenda Building*. Boston, MA: Allyn and Bacon.

Cockerham, Sean. 2014. "Fracking Drives Growth in Sand Mining, Raises New Health-Risk Questions." *McClatchy DC*, October 13. www.mcclatchydc.com/news/nation-world/national/economy/article24774538.html.

Coin, Glenn. 2015. "New York State Officially Bans Fracking." The Post-Standard, June 29. http://www.syracuse.com/news/index.ssf/2015/06/new_york_officially_bans_hydrofracking.html

Colaneri, Kate. 2014. "Wolf: New York's Fracking Ban Is 'Unfortunate'." *National Public Radio*, December 18. https://stateimpact.npr.org/pennsylvania/2014/12/18/wolf-new-yorks-fracking-ban-is-unfortunate/.

Colburn, Theo, Carol Kwiatkowski, Kim Schultz, and Mary Bachran. 2011. "Natural Gas Operations from a Public Health Perspective." *Human and Ecological Risk Assessment: An International Journal* 17(5): 1039–1056.

Colorado Community Development Office. 2013. "Local Government Land Use Authority in Colorado." www.colorado.gov/cs/Satellite?blobcol=urldata&blobheadername1=Content-Disposition&blobheadername2=Content-Type&blobheadervalue1=inline%3B+filename%3D%22Land+Use+Planning+in+Colorado.pdf%22&blobheadervalue2=application%2Fpdf&blobkey=id&blobtable=MungoBlobs&blobwhere=1251883675536&ssbinary=true (January 2, 2014).

Colorado Department of Local Affairs. 2015. "Direct Distribution—Counties and Municipalities." www.colorado.gov/pacific/dola/direct-distribution-counties-municipalities.

———. 2015a. "Direct Distribution Payments." www.colorado.gov/pacific/dola/direct-distribution-payments.

Colorado Department of Natural Resources. 2014. "CDPHE's Regulation of Oil and Gas Development." http://dnr.state.co.us/ogtaskforce/Documents/CDPHE%20Regulation%20of%20Oil%20and%20Gas102014.pdf.

Colorado Department of Public Health and Environment. 2014. "Oil and Gas Industry Environmental Requirements." www.colorado.gov/cs/Satellite/CDPHE-AP/CBON/1251596441882.

———. N.d. "Frequently Asked Questions." www.colorado.gov/pacific/sites/default/files/DC_Env_Oil-and-gas_FAQ.pdf.

Colorado Division of Water Resources. 2015. "Water Sources and Demand for the Hydraulic Fracturing of Oil and Gas Wells in Colorado from 2010 through 2015." http://water.state.co.us/DWRIPub/CGWC%20Meetings%20and%20Process%20Documents/Oil%20and%20Gas%20Water%20Sources%20Fact%20Sheet%20-%20Final.pdf.

Colorado Legislative Council Staff. 2013. "Oil and Gas Severance Tax Information." www.coloradopetroleumassociation.org/wp-content/uploads/2014/04/Oil-and-Gas-Severance-Taxes-IP.pdf.

———. 2010. "Colorado Mineral Tax Revenue." www.colorado.gov/cs/Satellite?blobcol=urldata&blobheader=application%2Fpdf&blobkey=id&blobtable=MungoBlobs&blobwhere=1251672450913&ssbinary=true.

———. 2008. "Severance Tax and FML Revenue." www.colorado.gov/pacific/sites/default/files/Allocation%20of%20Severance%20Taxes.pdf.

Colorado Municipal League. N.d. "Memorandum of Understanding." www.cml.org/uploadedFiles/CML_Site_Map/_Global/training/cle_thur_mou_arapahoe.pdf.

Colorado Oil and Gas Association. 2012. "Hydraulic Fracturing Whitepaper." www.coga.org/pdfs_facts/hfwhitepaper.pdf

———. 2012a. "The Reality of Increased Setbacks." www.coga.org/index.php/Newsroom/FactSheetsArticle/the_reality_of_increased_setbacks#sthash.4uJnOPTe.QEXL2ek5.dpbs.

———. 2012b. "The Role of Local Government in Oil and Gas Regulation." www.coga.org/pdfs_policy/Local_Government_Whitepaper.pdf (August 3, 2013).

———. 2012c. "Total State Severance Tax Revenue." www.coga.org/pdf_Basics/Basics_SeveranceTax.pdf

———. 2011. "Colorado Oil & Gas Industry Tax Whitepaper." www.naro-us.org/Resources/Documents/COGA%20Whitepaper.pdf.

Colorado Oil and Gas Conservation Commission. 2012. "Hydraulic Fracturing in La Plata County & Colorado La Plata County League of Women's Voters." www.lwvlaplata.org/files/lpco_lwv_march3-12-12_resized_part1.pdf.

Colorado Secretary of State. 2014. "Total Registered Voter by Status." www.sos.state.co.us/pubs/elections/VoterRegNumbers/2014/January/VoterCountsByStatus.pdf.

Connor, John A., Kenneth Walker, Lisa Molofsky, Ernesto Baca, Elaine Gie, and Thomas McHugh. 2016. "Comment on Impact to Underground Sources of Drinking Water and Domestic Wells from Production Well Stimulation and Completion Practices in the Pavillion, Wyoming, Field." *Environmental Science & Technology* 50(19): 10769–10770.

Cook, Jeffrey J. 2015. "Let's Make a Deal: Unraveling Stakeholder Influence on Colorado Fracking Policy." *Journal of Natural Resources Policy Research* 7(4): 267–285.

CRS Reports & Analysis. 2015. "Federal Court Rules That Bureau of Land Management Likely Lacks Authority to Promulgate Fracking Rule." www.fas.org/sgp/crs/misc/blmfracking.pdf.

Csoka, Louis V. 2007. "The Dream of Greater Municipal Autonomy: Should the Legislature or the Courts Modify Dillon's Rule, a Common Law Restraint on Municipal Power?" *North Carolina Central Law Journal* 29: 194–223.

Cusick, Marie. 2014. "Scientists Document Gas Development Disturbing Forest Ecosystems." *National Public Radio*, January 9. "https://stateimpact.npr.org/pennsylvania/2014/01/09/scientists-document-gas-development-disturbing-forest-ecosystems/.

Dahl, Robert. 1961. *Who Governs? Democracy and Power in an American City.* New Haven, CT: Yale University Press.

Dalbey, Beth. 2014. "Fracking Moratoriums Idle Drilling Equipment—At Least Temporarily: Rochester Hills Is the Latest of Several Municipalities Getting Moratoriums on Gas and Oil Exploration as State Considers Rules." *West-Bloomfield Patch*, August 28. http://patch.com/michigan/westbloomfield/fracking-moratoriums-idle-drilling-equipment-least-temporarily.

Daley, Dorothy M., and James C. Garand. 2005. "Horizontal Diffusion, Vertical Diffusion and Internal Pressure in State Environmental Policymaking, 1989–1998." *American Politics Research* 33(5): 615–644.

———, and David F. Layton. 2004. "Policy Implementation and the Environmental Protection Agency: What Factors Influence Remediation at Superfund Sites?" *Policy Studies Journal* 32(3): 375–392.

Davenport, Coral. 2016. "E.P.A. Methane Leak Rules Take Aim at Climate Change." *New York Times*, May 12. www.nytimes.com/2016/05/13/us/obama-methane-epa.html.

Davis, Charles. 2014. "Substate Federalism and Fracking Policies: Does State Regulatory Authority Trump Local Land Use Autonomy?" *Environmental Science & Technology* 48(15): 8397-8403.

———. 2012. "The Politics of 'Fracking': Regulating Natural Gas Drilling Practices in Colorado and Texas." *Review of Policy Research* 29(2): 177–191.

———, and Jonathan M. Fisk. 2017. "Mitigating Risks from Fracking-Related Earthquakes: Assessing State Regulatory Decisions. *Forthcoming at Society and Natural Resources*. http://dx.doi.org/10.1080/08941920.2016.1273415.

———. 2014. "Analyzing Public Support for Fracking in the U.S." *Review of Policy Research* 31(1): 1–16.

Davis, Mary E. 2005. "Environmental Politics in the US: A Study of State Sulfur Dioxide Standards." *Journal of Environmental Assessment Policy and Management* 7(3): 331–354.

De Avila, Joseph, Mike Vilensky, and Russell Gold. 2014. "New York Communities Can Ban Fracking, Court Rules Opponents Have Used Zoning Laws in More Than 170 Municipalities to Outlaw the Practice." *Wall Street Journal*, June 30. www.wsj.com/articles/new-york-towns-and-cities-can-ban-fracking-court-rules-1404145435.

Del Valle, Fernando. 2015. "Local Protesters Rally Outside Private LNG Forum, Demonstration." *The Monitor*, October 20. www.themonitor.com/news/local/local-protesters-rally-outside-private-lng-forum-demonstration/article_02fa4770-7792-11e5-b493–9bae8a365fca.html.

Dennis, Justin. 2015. "Ballot Initiatives off Table for Fracking Opponents." *Star Beacon*, September 4. www.starbeacon.com/news/local_news/ballot-initiatives-off-table-for-fracking-opponents/article_53c4432f-f6fe-5c53-b760-086e5efba0a1.html.

———. 2015a. "Rejecting Injection Wells: Officials Seek Increased Local Control of Fracking Industry." *Akron Star-Beacon*, July 28. www.starbeacon.com/news/rejecting-injection-wells-officials-seek-increased-local-control-of-fracking/article_ea1fbe40-3599–11e5-a130-c7d4df0a9111.html.

Denver Post Editorial Board. 2015. "Time to De-Bruce Colorado Severance Taxes?" *The Denver Post*, March 30. www.denverpost.com/editorials/ci_27801148/time-de-bruce-colorado-severance-taxes.

———. 2014. "Boulder District Judge Tosses Lafayette's Fracking Ban." *The Denver Post*, August 27. www.denverpost.com/business/ci_26419597/boulder-district-judge-tosses-lafayettes-fracking-ban.

DePillis, Lydia. 2015. "The Boom-Proof Economy: How to Handle a Fracking Bust." *Washington Post*, January 15. www.washingtonpost.com/news/storyline/wp/2015/01/15/the-boom-proof-economy-how-to-handle-a-fracking-bust/.

Desvousges, William, V. Kerry Smith, and Hillery H. Rink, III. 1992. "Communicating Radon Risks Effectively: The Maryland Experience." *Journal of Public Policy and Marketing* 11(2): 68–78.

Detrow, Scott. 2013. "Five Years Into Fracking Boom, One Pa. Town at a Turning Point." *National Public Radio*, January 14. www.npr.org/2013/01/14/169363299/five-years-into-fracking-boom-one-pa-town-at-a-turning-point.

———. 2011. "Pennsylvania Firefighters Prepare For 'The Big One'." *National Public Radio*, November 1. https://stateimpact.npr.org/pennsylvania/2011/11/01/pennsylvania-firefighters-prepare-for-the-big-one/.

DeWitt, David. 2013. "Fracking Bans Have Mixed Success in Other Ohio Cities." *The Athens News*, November 13. www.athensnews.com/ohio/article-41113-fracking-bans-have-mixed-success-in-other-ohio-cities.html.

DiGiulio, Dominic C., and Robert B. Jackson. 2016. "Impact to Underground Sources of Drinking Water and Domestic Wells from Production Well Stimulation and Completion Practices in the Pavillion, Wyoming, Field." *Environmental Science & Technology* 50(8): 4524–4536.

DiPasquale, Denise, and Edward L. Glaeser. 1999. "Incentives and Social Capital: Are Homeowners Better Citizens?" *Journal of Urban Economics* 45(2): 354–384.

Dlouhy, Jennifer A. 2011. "Natural Gas Boom Projected to Fuel Job Growth." *Houston Chronicle*, December 5. www.chron.com/business/article/Natural-gas-boom-projected-to-fuel-job-growth-2347054.php.

Dolan, Matthew. 2015. "Ohio Gov. John Kasich to Propose $500 Million Tax Cut Plan Includes Cuts to Personal Income Tax, Increases Cigarette Tax." *Wall Street Journal*, February 1. www.wsj.com/articles/ohio-gov-john-kasich-to-propose-500-million-tax-cut-1422840665.

Downing, Bob. 2015. "Local Officials Seek More Local Control from Ohio on Drilling." *Akron Beacon-Journal*, July 9. www.ohio.com/blogs/drilling/ohio-utica-shale-1.291290/local-officials-seek-more-local-control-from-ohio-on-drilling-1.606723.

———. 2014. "Ohio Supreme Court Hears Arguments in Munroe Falls Drilling Case." *Akron Beacon-Journal*, February 26. www.ohio.com/blogs/drilling/ohio-utica-shale-1.291290/ohio-supreme-court-hears-arguments-in-munroe-falls-drlling-case-1.469390.

———. 2013. "Fracking Has Not Caused Water-Well Problems in Ohio." *Akron-Beacon Journal*, October 19. www.ohio.com/blogs/drilling/ohio-utica-shale-1.291290/fracking-has-not-caused-water-well-problems-in-ohio-1.437864.

———. 2013a. "Questions Remain in Youngstown Dumping Case, Hagan Says." *Akron Beacon-Journal*, February 28. www.ohio.com/blogs/drilling/ohio-utica-shale-1.291290/antero-seeking-more-barnesville-leases-in-belmont-county-1.371880?month=1&year=2013.

———. 2012. "Cincinnati Bans Injection Wells, an Ohio First." *Akron Beacon-Journal*, August 2. http://www.ohio.com/blogs/drilling/ohio-utica-shale-1.291290/cincinnati-bans-injection-wells-an-ohio-first-1.324187.

Dropkin, Alex. 2014. "What a Ban on Fracking in Denton Could Mean for the Rest of Texas." *National Public Radio*, April 8. http://stateimpact.npr.org/texas/2014/04/08/what-a-fracking-ban-in-denton-could-mean-for-texas/.

———, and Terrence Henry. 2015. "How the Denton Fracking Ban Could Work." *National Public Radio*, N.d. https://stateimpact.npr.org/texas/tag/denton/.

Dryzek, John. 2005. *The Politics of the Earth: Environmental Discourses*. 2nd ed. Oxford: Oxford University Press.

Duffy, Robert J. 2005. "Political Mobilization, Venue Change, and the Coal Bed Methane Conflict in Montana and Wyoming." *Natural Resources Journal* 45(2): 409–440.

Dumcius, Gintautas, 2016. "Massachusetts Senate OKs a Ban on New Pipelines within 1,000 Feet of a School or Senior Center." *The Republican*, May 26. www.masslive.com/politics/index.ssf/2016/05/massachusetts_senate_oks_a_ban.html.

Dunn, Sharon. 2014. "Fracking 101: Breaking Down the Most Important Part of Today's Oil, Gas Drilling." *Greeley Tribune*, January 5. www.greeleytribune.com/news/9558384-113/drilling-oil-equipment-wellbore.

Duroy, Quentin M. H. 2008. "Testing the Affluence Hypothesis: A Cross-Cultural Analysis of the Determinants of Environmental Actions." *The Social Science Journal* 45(3): 419–439.

Dyson, Cathy. 2015. "Attorney General: Localities Can Ban Fracking: Herring Reverses Stand of Predecessor, Says Zoning, Land-Use Laws Can Prohibit Drilling." *Fredericksburg Free Lance-Star*, May 9. www.fredericksburg.com/news/local/attorney-general-localities-can-ban-fracking/article_c4c13fb3-0619-5753-b6c8-a796b6af3525.html.

Eaton, Joe. 2013. "Results Mixed on Colorado and Ohio Fracking Ban Initiatives: Recount Is Likely Ahead for Broomfield, as Moratorium Fails by 13 Votes." *National Geographic*, November 6. http://news.nationalgeographic.com/news/energy/2013/11/131105-colorado-ohio-fracking-ban-election-results/.

Efstathiou, Jim, Jr. 2012. "Taxpayers Pay as Fracking Trucks Overwhelm Rural Cow Paths." *Bloomberg News*, May 15. www.businessweek.com/news/2012-05-15/taxpayers-pay-as-fracking-trucks-overwhelm-rural-cow-paths.

Eggers, William D., and John O'Leary. 1995. *Revolution at the Roots: Making Our Governments Smaller, Better, and Closer to Home*. New York: Free Press.

EIA. 2016. "Greenhouse Gas Emissions and Atmospheric Concentrations Have Increased Over the Past 150 Years." www.eia.gov/energyexplained/index.cfm?page=environment_how_ghg_affect_climate.

———. 2016a. "Natural Gas End Use." www.eia.gov/dnav/ng/ng_cons_sum_dcu_nus_a.htm.

———. 2016b. "Reserves Supply." www.eia.gov/state/data.cfm?sid=CO#Reserves Supply.

———. 2016c. "Reserves Supply." www.eia.gov/state/data.cfm?sid=OH#Reserves Supply.

———. 2016d. "Reserves Supply." www.eia.gov/state/data.cfm?sid=TX#Reserves Supply.

———. 2016e. "Short-Term Energy and Summer Fuel Outlook." www.eia.gov/forecasts/steo/.

———. 2016f. "State Inflow Capacity." www.eia.gov/naturalgas/pipelines/EIA-StatetoStateCapacity.xls.

———. 2015. "Sales of Fossil Fuels Produced from Federal and Indian Lands, FY 2003 through FY 2014." www.eia.gov/analysis/requests/federallands/pdf/eia-federallandsales.pdf.

———. 2015a. "U.S. Natural Gas Regulatory Authorities." www.eia.gov/pub/oil_gas/natural_gas/analysis_publications/ngpipeline/regulatory.html#domestic.

———. 2015b. "Today in Energy." www.eia.gov/todayinenergy/detail.cfm?id=20592#tabs_Slider-5.

———. 2014. "Natural Gas Producing Wells." www.eia.gov/dnav/ng/ng_prod_wells_s1_a.htm.
———. 2014a. "Natural Gas Production." www.eia.gov/dnav/ng/ng_prod_sum_dcu_stx_a.htm.
———. 2014b. "Natural Gas Gross Withdrawals and Production." www.eia.gov/dnav/ng/ng_prod_sum_dcu_NUS_a.htm.
———. 2014c. "Production of Crude Oil." www.eia.gov/dnav/pet/hist/LeafHandler.ashx?n=PET&s=MCRFPUS1&f=A.
———. 2013. "The Past 150 Years Show an Increase in Greenhouse Gases." www.eia.gov/energyexplained/index.cfm?page=environment_how_ghg_affect_climate.
———. 2013a. "Recoverable Shale Oil and Shale Gas Resources: An Assessment of 137 Shale Formations in 41 Countries Outside the United States." September 30. www.eia.gov/analysis/studies/worldshalegas/pdf/fullreport.pdf.
———. 2013b. "Today in Energy: Rail Delivery of U.S. Oil and Petroleum Products Continues to Increase, but Pace Slows." www.eia.gov/todayinenergy/detail.php?id=12031.
———. 2009. "Gas Production in Conventional Fields in the Lower 48." April 8. www.eia.gov/oil_gas/rpd/conventional_gas.jpg.
———. 2008. "About U.S. Natural Gas Pipelines—Transporting Natural Gas." www.eia.gov/pub/oil_gas/natural_gas/analysis_publications/ngpipeline/interstate.html.
———. 2005. "Overview of U.S. Legislation and Regulations Affecting Offshore Natural Gas and Oil Activity." www.eia.gov/pub/oil_gas/natural_gas/feature_articles/2005/offshore/offshore.pdf.
Elliott, Dan. 2015. "Oil-Gas Panel Suggests Consulting Role for Local Government." *Washington Times*, February 24. www.washingtontimes.com/news/2015/feb/24/colorado-oil-gas-debate-resumes-as-task-force-fini/?page=all.
Ellsworth, William L. 2013. "Injection-Induced Earthquakes." *Science* 341(6142): 142–150.
Elmore, Richard. 1980. "Backward Mapping: Implementation Research and Policy Decisions." *Political Science Quarterly* 94(4): 601–617.
Encana. 2013. "Completed Well Pads and Equipment." www.encana.com/pdf/communities/usa/CompletedWellPadsandEquipment(DJ).pdf.
Environmental Protection Agency [EPA]. 2016. "Clean Power Plan." www.epa.gov/cleanpowerplan/fact-sheet-components-clean-power-plan.
———. 2016a. "Hydraulic Fracturing for Oil and Gas: Impacts from the Hydraulic Fracturing Water Cycle on Drinking Water Resources in the United States, Main Report & Appendices (EPA/600/R-16/236)." www.epa.gov/hfstudy/executive-summary-hydraulic-fracturing-study-final-assessment-2016.
———. 2015. "The Hydraulic Fracturing Water Cycle." www2.epa.gov/hfstudy/hydraulic-fracturing-water-cycle#ftn6.
———. 2011. "Hydraulic Fracturing." www.epa.gov/hydraulicfracturing/process.html.
———. 2011a. "What Is Natural Gas Star?" www3.epa.gov/gasstar/documents/ngstar_mktg-factsheet.pdf.
———. 2010. "Summary of the Clean Water Act: 33 U.S.C. §1251 et seq. (1972)." www.epa.gov/lawsregs/laws/cwa.html.
Environmental Working Group. 2013. "Free Pass for Oil and Gas: Oil and Gas Industry Exemptions." www.ewg.org/research/free-pass-oil-and-gas/oil-and-gas-industry-exemptions.

Estabrook, Rachel. 2015. "Hickenlooper: Oil and Gas Task Force Recs 'Probably Should be Enacted'." *Colorado Public Radio*, February 26. www.cpr.org/news/story/hickenlooper-oil-and-gas-task-force-recs-probably-should-be-enacted.

Federal Wildlife Service. N.d. "Contaminant Issues—Oil Field Waste Pits." www.fws.gov/mountain-prairie/contaminants/contaminants1a.html.

Fehling, Dave. 2014. "The Gas Well Next Door: How Drilling Changed in Fort Worth." *National Public Radio*, March 11. http://stateimpact.npr.org/texas/2014/03/11/how-drilling-changed-fort-worth/.

———. 2012a. "Fracking's Link to Smog Worries Some Texas Cities." *National Public Radio*, August 7. http://stateimpact.npr.org/texas/2012/08/07/frackings-link-to-smog-worries-some-texas-cities/.

———. 2012b. "Texas Drillers get Big Tax Breaks." *National Public Radio*, January 9. https://stateimpact.npr.org/texas/2012/01/09/texas-drillers-get-big-tax-breaks/.

Feiock, Richard C., and Jae-Hoon Kim. 2001. "Form of Government, Administrative Organization and Economic Development Policy." *Journal of Public Administration Research and Theory* 11(1): 29–49.

———, and Christopher Stream. 2001. "Environmental Protection Versus Economic Development: A False Trade-Off?" *Public Administration Review* 61(3): 313–321.

Feldman, Martha S., and Anne M. Khademian. 2002. "To Manage Is to Govern." *Public Administration Review* 62(5): 541–554.

Finley, Bruce. 2015. "Federal Judge Blocks BLM Rules for Fracking on Public Lands." *The Denver Post*, September 30. www.denverpost.com/news/ci_28901828/federal-judge-blocks-blm-rules-fracking-public-lands.

———. 2014. "Scientists Flying over Colorado Oil Boom Find Worse Air Pollution." *The Denver Post*, May 8. www.denverpost.com/News/Local/ci_25719742/Scientists-flying-over-Colorado-oil-boom.

———. 2013. "Drilling in Gung-Ho Greeley, Hits Opposition Near Westside Homes." *The Denver Post*, May 5. www.denverpost.com/ci_23174358/drilling-gung-ho-greeley-hits-opposition-near-west.

———. 2012. "Fracking Bidders Top Farmers at Water Auction." *The Denver Post*, April 2. www.denverpost.com/entertainmentcolumnists/ci_20306480/fracking-bidders-top-farmers-at-water-auction.

Fiorino, Daniel J. 2006. *The New Environmental Regulation*. Cambridge: MIT Press.

———. 2001. "Environmental Policy as Learning: A New View of an Old Landscape." *Public Administration Review* 61(3): 322–334.

Fisk, Jonathan M. 2016. "Fractured Relationships: Exploring Municipal Defiance in Colorado, Texas, and Ohio." *State and Local Government Review* 48(2): 1–12.

———. 2013. "The Right to Know? State Politics of Fracking Disclosure." *Review of Policy Research* 30(4): 345–365.

Fitzpatrick, David, and Drew Griffin. 2015. "Oil Tankers: Danger on the Rails." *CNN*, May 14. www.cnn.com/2015/05/14/us/oil-tank-investigation/.

Fontenot, Brian E., Laura R. Hunt, Zacariah L. Hildenbrand, Doug D. Carlton, Jr., Hyppolite Oka, Jayme L. Walton, Dan Hopkins, Alexandra Osorio, Bryan Bjomdal, Qinhong H. Hu, and Kevin A. Schug. 2013. "An Evaluation of Water Quality in Private Drinking Water Wells Near Natural Gas Extraction Sites in the Barnett Shale Formation." *Environmental Science & Technology* 47(17): 10032–10040.

Foran, Clare. 2014. "How Many Jobs Does Fracking Really Create?" *National Journal*, April 14. www.nationaljournal.com/new-energy-paradigm/how-many-jobs-does-fracking-really-create-20140414.

———. 2013. "House Passes Bill to Block Federal Fracking Regulation." *National Journal*, November 21. www.nationaljournal.com/energy/house-passes-bill-to-block-federal-fracking-regulation-20131121.

Fort Worth Chamber of Commerce. 2014. "The Economic and Fiscal Contribution of the Barnett Shale: Impact of Oil and Gas Exploration and Production on Business Activity and Tax Receipts in the Region and State." www.fortworthchamber.com/chamber/docs/BSISeptember2014.pdf.

FracFocus. 2016. "Public Disclosure." https://fracfocus.org/chemical-use/chemicals-public-disclosure.

———. 2015. "Hydraulic Fracturing: How It Works?" https://fracfocus.org/hydraulic-fracturing-how-it-works/history-hydraulic-fracturing.

Frosch, Dan. 2013. "Wyoming May Act to Plug Abandoned Wells as Natural Gas Boom Ends." *New York Times*, December 24. www.nytimes.com/2013/12/25/us/state-may-act-to-plug-abandoned-wyoming-wells-as-natural-gas-boom-ends.html?_r=0.

———, and Russell Gold. 2015. "How 'Orphan' Wells Leave States Holding the Cleanup Bag." *Wall Street Journal*, February 25. www.wsj.com/articles/how-orphan-wells-leave-states-holding-the-cleanup-bag-1424921403.

Fry, Leah. 2013. "Niles City Council Repeals Ordinance Banning Fracking." *NBC-WFMJ*, September 25. www.wfmj.com/story/23475330/niles-city-council-repeals-ordinance-banning-fracking.

Funk, John. 2016. "Ohio Gas and Oil Production Soaring, Fewer Drilling Rigs at Work." *Cleveland Plain Dealer*, March 10. www.cleveland.com/business/index.ssf/2016/03/ohio_gas_and_oil_production_so.html.

———. 2014. "Ohio Gov. Kasich Would Collect $874 Million from Shale Wells in Next Three Years." *Cleveland Plain-Dealer*, March 11. www.cleveland.com/business/index.ssf/2014/03/ohio_gov_kasich_would_collect.html.

Gabriel, Trip. 2014. "In Compromise Plan, Limited Fracking Is Approved for National Forest in Virginia." *New York Times*, November 18. www.nytimes.com/2014/11/19/us/us-backs-limited-use-of-oil-drilling-technique-in-national-forest-compromise.html?_r=0.

Gainsborough, Juliet. 2003. "To Devolve or Not to Devolve? Welfare Reform in the States." *Policy Studies Journal* 31(4): 603–624.

Gang, Duane. 2013. "Friction over Fracking Grows in South." *USA Today*, September 30. www.usatoday.com/story/news/nation/2013/09/30/fracking-south-natural-gas-drilling-tennessee/2897895/.

Garfield, Andrew. 2015. "Energy Takes Center Stage." *Philadelphia Inquirer*, March 9. www.philly.com/philly/opinion/inquirer/20150309_Energy_takes_center_stage.html.

———. 2015a. "U.S. Energy Policies Having Devastating Effect on Anti-Western Regimes." *Milwaukee Journal-Sentinel*, March 9. www.jsonline.com/news/opinion/us-energy-policies-having-devastating-effect-on-anti-western-regimes-b99458779z1-295682871.html.

Garrett, Robert. 2015. "Texas' Rainy Day Fund Overflows—and Divides Legislators." *Dallas Morning News*, March 20. www.dallasnews.com/news/politics/state-politics/20150320-texas-rainy-day-fund-overflows-and-divides-legislators.ece.

Gazette Editorial Board. 2014. "Frac Sand Mining: Its Fate in NE Iowa Should Rest in Local Hands." *The Gazette*, February 16. http://thegazette.com/2014/02/16/frac-sand-mining-its-fate-in-ne-iowa-should-rest-in-local-hands/#ixzz30Us3af5r (March 14, 2014).

Gearino, Dan. 2015. "Ohio's Fracking Boom Hits Speed Bump." *The Columbus Dispatch*, November 27. www.dispatch.com/content/stories/business/2015/11/27/fracking-boom-hits-speed-bump.html.

Geewax, Marilyn. 2013. "What's Under Youngstown May Help What's On Top." *National Public Radio*, May 29. www.npr.org/2013/05/29/185804016/whats-under-youngstown-may-help-whats-on-top.

Getches-Wilkinson Center for Natural Resources, Energy, and the Environment 2013 at the University of Colorado School of Law. 2013. "Intermountain Oil and Gas BMP Project." www.oilandgasbmps.org/laws/colorado_law.php.

Goad, Benjamin. 2014. "Oil Lobby: Fracking Saved Americans $248 billion Last Year." *The Hill*, October 30. http://thehill.com/regulation/energy-environment/222335-oil-lobby-fracking-saved-americans-248-billion-last-year.

Goggin, Malcolm L., Ann Bowman, James Lester, and Laurence O'Toole. 1990. *Implementation Theory and Practice: Toward a Third Generation*. New York: Scott Foresman.

Goho, Shaun. 2012. "Municipalities and Hydraulic Fracturing: Trends in State Preemption." *Planning and Environmental Law* 64(7): 3–9.

Gold, Russell. 2015. "Texas Prohibits Local Fracking Bans." *Wall Street Journal*, May 18. www.wsj.com/articles/texas-moves-to-prohibit-local-fracking-bans-1431967882.

———, and Tom McGinty. 2013. "Energy Boom Puts Wells in America's Backyards; Hydraulic Fracturing Largely Driving Transformation of the Nation's Landscape." *Wall Street Journal*, October 25. http://online.wsj.com/news/articles/SB10001424052702303672404579149432365326304.

Golden, John M., and Hannah Wiseman. 2015. "The Fracking Revolution: Shale Gas as a Case Study in Innovation Policy." *Emory Law Journal* 64(3): 957–1037.

Goldenberg, Suzanne. 2014. "Fracking Is Depleting Water Supplies in America's Driest Areas, Report Shows." *The Guardian*, February 5. www.theguardian.com/environment/2014/feb/05/fracking-water-america-drought-oil-gas.

———. 2013. "A Texan Tragedy: Ample Oil, No Water: Fracking Boom Sucks Away Precious Water from Beneath the Ground, Leaving Cattle Dead, Farms Bone-Dry and People Thirsty." *The Guardian*, August 11. www.theguardian.com/environment/2013/aug/11/texas-tragedy-ample-oil-no-water.

Goodwyn, Wade. 2015. "New Texas Law Makes Local Fracking Bans Illegal." *National Public Radio*, May 22. www.npr.org/2015/05/20/408156948/new-texas-law-makes-local-fracking-bans-illegal.

Gopalakrishnan, Sathya, and H. Allen Klaiber. 2014. "Is the Shale Energy Boom a Bust for Nearby Residents? Evidence from Housing Values in Pennsylvania." *American Journal of Agricultural Economics* 96(1): 43–66.

Gorman, Joe. 2013. "Trumbull County Makes Deal for Water: Being Used for North Jackson Well Site." *Tribune Chronicle*, February 8. www.tribunechronicle.com/page/content.detail/id/582996/Trumbull-County-makes-deal-for-water.html?nav=5003.

Gotherman, John. 2002. "Municipal Home Rule and Charters." www.columbus.gov/uploadedFiles/Columbus/Elected_Officials/City_Council/

Charter_Review_Commission/Ohio%20Municipal%20Home%20Rule%20 Charters,%20John%20E.%20Gotherman.pdf.
Grand Junction Staff. 2015. "Club 20 Stands Firm on Severance Tax: Organization Opposes Diverting Revenue from Mineral Sources to Offset Effects of TABOR." *Grand Junction Sentinel*, January 26. www.gjsentinel.com/news/articles/club-20-stands-firm-on-severance-tax.
Grantham, Joseph, and Colorado Division of Water Resources. 2011. "A Synopsis of Colorado Water Law." http://water.state.co.us/DWRIPub/DWR%20 General%20Documents/SynopsisofCOWaterLaw.pdf.
Gratton, Lynda, and Tamara J. Erikson. 2007. "Eight Ways to Build Collaborative Teams." *Harvard Business Review* 11(November): 1–11.
Gronewold, Nathaniel. 2015. "More Jobs, Rigs Shed Ahead of Third-Quarter Reports." *Environment and Energy Publishing*, October 19. www.eenews.net/energywire/2015/10/19/stories/1060026505.
———. 2015a. "Rocky Mountain Goodbye: Encana Exits Colo." *Environment and Energy Publishing*, October 9. www.eenews.net/energywire/2015/10/09/stories/1060026112.
Guardian Staff. 2015. "Texas Governor Bans Cities and Towns from Banning Fracking Greg Abbott Ensures State Legislature Has Sole Authority over Oil and Gas Regulation in Texas." *The Guardian*, May 18. www.theguardian.com/us-news/2015/may/19/fracking-texas-governor-stops-cities-and-towns-banning-hydraulic-gas-mining.
Gutiérrez, Bertrand M. 2015. "N.C. Lawmakers Counter Local Fracking Bans with Last-Minute Legislative Action." *Winston-Salem Journal*, October 5. www.journalnow.com/news/local/n-c-lawmakers-counter-local-fracking-bans-with-last-minute/article_1a3d60a8-5876–577e-8a21–93ccf15bcc34.html.
Haluszczak, Lara O., Arthur W. Rose, and Lee R. Kump. 2012. "Geochemical Evaluation of Flowback Brine from Marcellus Gas Wells in Pennsylvania, USA." *Applied Geochemistry* 28(January): 55–61.
Hamilton, James T. 1995. "Testing for Environmental Racism: Prejudice, Profits, Political Power?" *Journal of Policy Analysis and Management* 14(1): 107–132.
———, and W. Kip Viscusi. 1999. *Calculating Risks? The Spatial and Political Dimensions of Hazardous Waste Policy*. Cambridge: MIT Press.
Handy, Ryan M. 2014. "Judge Overturn Fort Collins Fracking Moratorium." *The Coloradoan*, August 7. www.coloradoan.com/story/news/local/2014/08/07/judge-overturns-fort-collins-fracking-moratorium/13743031/.
Haque, Asim. 2015. "Written Remarks of Asim Z. Haque." www.ferc.gov/CalendarFiles/20150602092302-Haque,%20PUCO.pdf.
Hardwick, Reginald, and Mark Schnyder. 2013. "Water Restrictions Begin for Millions of North Texans." *NBCDFW*. www.nbcdfw.com/news/local/Water-Restrictions-Begin-For-Millions-of-North-Texans-209690921.html.
Harper, John. 2015. "Akron's Drinking Water Supply Surrounded by Oil Wells, a Cash Cow for the City." *Akron Beacon-Journal*, October 1. www.cleveland.com/akron/index.ssf/2015/10/akrons_drinking_water_supply_i.html.
Hartman, Kristyn. 2014. "Eastern Ohio City Getting New Life with New Money from Fracking." www.10tv.com/content/stories/2014/05/15/cambridge-fracking-turns-eastern-ohio-city-to-boomtown.html.

Hays, Richard Allen. 1988. "State-Local Relations in Policy Implementation: The Case of Highway Transportation in Iowa." *Publius: The Journal of Federalism* 18(1): 79–95.

Haythorn, Russell. 2013. "Fracking and Its Effects on Public Health Still Debatable: Some Argue Fracking Is Making Them Sick." May 20. www.thedenverchannel.com/news/local-news/fracking-and-its-effects-on-public-health-still-debatable.

Healy, Jack. 2013. "Supporting Oil and Gas, but Resisting Encroachment." *New York Times*, June 9. www.nytimes.com/2013/06/10/us/supporting-oil-and-gas-but-resisting-encroachment.html?_r=0.

———. 2012. "For Farms in the West, Oil Wells Are Thirsty Rivals." *New York Times*, September 5. www.nytimes.com/2012/09/06/us/struggle-for-water-in-colorado-with-rise-in-fracking.html?pagewanted=all.

Heclo, Hugh. 1978. "Issue Networks and the Executive Establishment." In *The New American Political System*. ed. Anthony King. Washington, DC: American Enterprise Institute.

Heikkila, Tanya, Jonathan J. Pierce, Samuel Gallaher, Jennifer Kagan, Deserai A. Crow, and Christopher M. Weible. 2014. "Understanding a Period of Policy Change: The Case of Hydraulic Fracturing Disclosure Policy in Colorado." *Review of Policy Research* 31(2): 65–87.

Heinkel-Wolfe, Peggy. 2015. "Council Turns to Zoning Powers after Reducing Well Setbacks to 1,000 feet." *Denton Record-Chronicle*, August 4. www.dentonrc.com/local-news/local-news-headlines/20150804-council-turns-to-zoning-powers-after-reducing-well-setbacks-to-1000-feet.ece.

———. 2015a. "Protesters Picket Well Site in Denton Where Fracturing Has Resumed." *Denton Record-Chronicle*, May 27. www.dentonrc.com/local-news/local-news-headlines/20150527-fracking-battle-in-texas-not-over.ece.

———. 2015b. "Setbacks Likely to Tighten Up." *Denton Record-Chronicle*, July 27. www.dentonrc.com/local-news/local-news-headlines/20150727-setbacks-likely-to-tighten-up.ece.

———, and Lowell Brown. 2012. "Stuck in the Middle: Research: 'Socially Vulnerable' Areas Tend to Have Most Gas Wells Near Schools." *Denton Record-Chronicle*, February 11. www.dentonrc.com/local-news/local-news-headlines/20120211-stuck-in-the-middle.ece.

Hempel, Lamont. 2009. "Conceptual and Analytical Challenges in Building Sustainable Communities." In *Toward Sustainable Communities*. 2nd ed. eds. Daniel Mazmanian and Michael Kraft. Cambridge, MA: MIT Press.

Hendricks, David. 2016. "Texas Workforce Shows Signs of Weakness." *San Antonio Express-News*, April 15. www.expressnews.com/business/local/article/Texas-workforce-shows-signs-of-weakness-7251906.php.

Henry, Devin. 2015. "Oklahoma Blocks Local Fracking Bans." *The Hill*, June 1. http://thehill.com/policy/energy-environment/243645-oklahoma-blocks-local-fracking-bans.

Henry, Terrence. 2013. "Dallas City Council Denies Permits for Fracking." *National Public Radio*, August 29. http://stateimpact.npr.org/texas/2013/08/29/dallas-city-council-denies-permits-for-fracking/.

———. 2013a. "What We Know about Fracking Activity and the Ohio Earthquake." *National Public Radio*. https://stateimpact.npr.org/texas/2012/01/02/what-we-know-fracking-activity-and-the-ohio-earthquake/.

Henry, Tom. 2013b. "B.G. Council Unanimously Passes Ban on Fracking: City Ordinance's Strength is Untested." *Toledo Blade*, September 16. www.toledoblade.com/Energy/2013/09/17/B-G-council-unanimously-passes-ban-on-fracking.html#jQ7p49FwPCESzuWU.99/.

Hersher, Rebecca. 2016. "USGS Announces Largest Oil And Gas Deposit Ever Assessed in U.S." *National Public Radio*, November 17. www.npr.org/sections/thetwo-way/2016/11/16/502337471/usgs-announces-its-largest-oil-and-gas-discovery-ever.

Higgs, Robert. 2014. "Ohio Supreme Court Case Over Munroe Falls' Drilling Laws Could Impact Home Rule Authority across the State." *Cleveland Plain-Dealer*, February 25. www.cleveland.com/open/index.ssf/2014/02/ohio_supreme_court_case_over_m.html.

Hindi, Saja. 2013. "Fort Collins Residents to Vote on Fracking Moratorium: Proponents Cite Environmental Concerns: Opponents Cite Misinformation and Potential Litigation." *Loveland Reporter-Herald*, October 29. www.reporterherald.com/news/larimer-county/ci_24299370/fort-collins-residents-vote-fracking-moratorium.

Hobbs, Heidi H. 1994. *City Hall Goes Abroad: The Foreign Policy of Local Politics*. Thousand Oaks, CA: SAGE Publications.

Hodos, Jerome. 2009. "Against Exceptionalism: Intercurrence and Intergovernmental Relations in Britain and the United States." In *The City in American Political Development*. ed. Richardson Dilworth. New York: Routledge.

Hoover, Shane. 2015. "Service Companies Hit by Drilling Slowdown." *Canton Repository*, February 26. www.cantonrep.com/article/20150226/NEWS/150229401.

———. 2015a. "State: Fewer Horizontal Drilling Permits this Year." *Canton Repository*, April 8. www.cantonrep.com/article/20150408/NEWS/150409363/0/SEARCH/?Start=1.

Hopkins, Carol. 2013. "'Fracking' Town Halls to be Held throughout the Year." *Oakland Press*, September 3. www.theoaklandpress.com/general-news/20130303/fracking-town-halls-to-be-held-throughout-the-year.

Howarth, Robert W., Renee Santoro, and Anthony Ingraffea. 2011. "Methane and the Greenhouse-Gas Footprint of Natural Gas from Shale Formations." *Climatic Change* 106(4): 679–690.

Howell, Susan E., and Shirley B. Laska. 1992. "The Changing Face of the Environmental Coalition: A Research Note." *Environment and Behavior* 24(1): 134–144.

Hupe, Peter. 2011. "The Thesis of Incongruent Implementation: Revisiting Pressman and Wildavsky." *Public Policy and Administration* 26(1): 63–80.

Husted, Kristofor. 2014. "Grain Glut Clogs Railways, Rivers, Roads." *KCUR*, December 22. http://kcur.org/post/grain-glut-clogs-railways-rivers-roads#stream/0.

ICF International Associates. 2014. "U.S. Oil Impacts." www.api.org/~/media/Files/Policy/Hydraulic_Fracturing/ICF-Hydraulic-Fracturing-Oil-Impacts.pdf.

Institute of Government Solutions, and University of California Berkley. 2013. "Fracking's Effects on Ohio's Economy . . . For Now." http://brr.berkeley.edu/2013/09/frackings-effects-on-ohios-economy-for-now/.

Integra Realty Resources. 2010. "Flower Mound Well Site Impact Study." www.flower-mound.com/DocumentCenter/View/1456.

International Council of Chemical Associations. 2016. "FAQ." www.icca-chem.org/en/Home/Responsible-care/FAQs/.

Interstate Natural Gas Association of America. 2013. "Building Interstate Natural Gas Transmission Pipelines: A Primer." www.ingaa.org/file.aspx?id=19618.

Interstate Oil and Gas Compact Commission. 2016. "State Progress." http://groundwork.iogcc.ok.gov/topics-index/orphaned-wells/state-progress.

———. 2015. "Our History." http://iogcc.publishpath.com/history.

Ivey, Mike. 2013. "Wisconsin at 'Global Epicenter' of Frac Sand Mining Industry." *Madison Capital Times*, October 10. http://host.madison.com/news/local/writers/mike_ivey/wisconsin-at-global-epicenter-of-frac-sand-mining-industry/article_45690930-3125-11e3-ba86-0019bb2963f4.html#ixzz3TvBeo3UU.

Jackson, Robert B. 2016. "Shale Gas and Water Resources." February 14. https://aaas.confex.com/aaas/2016/webprogram/Paper16793.html.

———, Brooks R. Pearson, Stephen G. Osborn, Nathaniel R. Warner, and Avner Vengosh. 2011. *Research and Policy Recommendations for Hydraulic Fracturing and Shale Gas Extraction*. Durham, NC: Center on Global Change, Duke University.

———, Avner Vengosh, J. William Carey, Richard J. Davies, Thomas H. Darrah, Francis O'Sullivan, and Gabrielle Pétron. 2014. "The Environmental Costs and Benefits of Fracking." *Annual Review of Environment and Resources* 39: 327–362.

Jacquet, Jeffrey B. 2014. "Review of Risks to Communities from Shale Energy Development." *Environmental Science and Technology* 48(15): 8321–8333.

———. 2012. "Landowner Attitudes toward Natural Gas and Wind Farm Development in Northern Pennsylvania." *Energy Policy* 50(November): 677–688.

Jaffe, Mark. 2015. "Colorado Oil, Gas Task Force Sends 9 Measures on to Governor's Desk." *The Denver Post*, February 24. www.denverpost.com/business/ci_27592566/colorado-oil-gas-task-force-sends-9-measures.

———. 2014. "Colorado Legislative Committee OKs Oil and Gas Health Impact Study." *The Denver Post*, April 1. www.denverpost.com/business/ci_25468580/colorado-legislative-committee-oks-oil-and-gas-health#ixzz2y9i0vuNS.

———. 2014a. "Hickenlooper Compromise Keeps Oil and Gas Measures off Colorado Ballot." *The Denver Post*, August 5. www.denverpost.com/business/ci_26272493/hickenlooper-tries-broker-last-minute-deal-oil-gas-colorado.

———. 2014b. "Oil and Gas Companies Playing, but Pay to Colorado Counties is Uneven." *The Denver Post*, April 13. www.denverpost.com/business/ci_25551329/oil-and-gas-companies-playing-but-pay-colorado.

———. 2014c. "Researchers Assess Emissions from Colorado Oil and Gas Fields." *The Denver Post*, November 16. www.denverpost.com/business/ci_26942340/researchers-assess-emissions-from-colorado-oil-and-gas.

———. 2014d. "When Drought Occurs, Fracking and Farming Collide." *The Denver Post*, February 9. www.denverpost.com/business/ci_25089583/when-drought-occurs-fracking-and-farming-collide.

———. 2013. "Colorado Joins in Suit to Knock Down Longmont Fracking Ban." *The Denver Post*, July 11. www.denverpost.com/ci_23643679/state-joins-suit-knock-down-longmont-fracking-ban#ixzz2pIi6cAMQ.

———. 2013a. "Colorado Oil and Gas Well Setback Fight Headed to State Legislature." *The Denver Post*, February 11. www.denverpost.com/ci_22566155/colorado-oil-and-gas-commission-votes-500-foot.

———. 2012. "Gov. Hickenlooper Says Longmont Drilling Rules Must be Challenged." *The Denver Post*, August 15. www.denverpost.com/ci_21319130/gov-hickenlooper-says-longmont-drilling-rules-must-be#ixzz2pIdYlHMA.

Jean, Sheryl. 2015. "Texas Job Growth Will Slow Even More This Year." *Dallas Morning News*, January 14. www.dallasnews.com/business/headlines/20150113-texas-job-growth-will-slow-even-more-this-year.ece.

Jervis, Rick. 2014. "Oil! New Texas Boom Spawns Riches, Headaches." *USA Today*, January 15. www.usatoday.com/story/news/nation/2014/01/15/texas-oil-boom-fracking/4481977/.

John, DeWitt. 1994. *Civic Environmentalism: Alternatives to Regulation in States and Communities*. Washington, DC: CQ Press.

Johnson, Alan. 2014. "Ohio EPA Director Scott Nally Resigns Abruptly, Cryptically." *Columbus-Dispatch*, January 8. www.dispatch.com/content/stories/local/2014/01/07/Ohio-EPA-director-resigns.html.

Jones, Bryan D., and Frank R. Baumgartner. 2012. "From There to Here: Punctuated Equilibrium to the General Punctuation Thesis to a Theory of Government Information Processing." *Policy Studies Journal* 40(1): 1–20.

Jones, Dan. 2013. "Oil and Gas Cases Point to Challenges." *Northern Colorado Business Report*, April 5. www.ncbr.com/article/20130405/EDITION0806/130409954/-1/section/MULTIMEDIA03.

Jones, Robert Emmet, and Riley E. Dunlap. 1992. "The Social Bases of Environmental Concern: Have They Changed Over Time?" *Rural Sociology* 57(1): 28–47.

Kamienicki, Sheldon. 2006. *Corporate America and Environmental Policy: How Often Does Business Get Its Way?* Palo Alto, CA: Stanford University Press.

Kantor, Paul, and Stephen David. 1988. *The Dependent City: The Changing Political Economy of Urban Economic Development in the United States*. Glenview, IL. Scott Foresman and Company.

Karpus, Jennifer. 2010. "Broadview Heights being Mandatory Pooled into Two Gas and Oil Wells." *Cleveland Plain-Dealer*, August 13. www.cleveland.com/sunstarcourier/index.ssf/2010/08/broadview_heights_being_mandat.html.

Kasperowicz, Pete. 2013. "House Votes to Block Federal Fracking Rules." *The Hill*, November 20. http://thehill.com/blogs/floor-action/votes/190977-house-votes-to-block-federal-fracking-rules.

Kates, Robert, Thomas Parris, and Anthony Leiserowitz. 2005. "What Is Sustainable Development? Goals, Indicators, Values and Practice." *Environment* 47(3): 8–21.

Keller, Ann. 2009. *Science and Environmental Policy*. Cambridge, MA: MIT Press.

Kenney, Douglas S. 1999. "Historical and Sociopolitical Context of the Western Watersheds Movement." *JAWRA: Journal of the American Water Resources Association* 35(3): 493–503.

Keranen, Katie M., Matthew Weingarten, Geoffrey A. Abers, Barbara A. Bekins, and Shemin Ge. 2014. "Sharp Increase in Central Oklahoma Seismicity since 2008 Induced by Massive Wastewater Injection." *Science* 345(6195): 448–451.

Kettl, Donald F. 2005. *The Global Public Management Revolution*. Washington, DC: Brookings Institution Press.

———. 2002. *The Transformation of Governance: Public Administration for Twenty-First Century America*. Baltimore, MD: Johns Hopkins University Press.

Kiger, Patrick J. 2014. "Scientists Warn of Quake Risk from Fracking Operations." *National Geographic*, May 2. http://news.nationalgeographic.com/news/energy/2014/05/140502-scientists-warn-of-quake-risk-from-fracking-operations/.

Kincaid, John. 1999. "The International Competence of US States and Their Local Governments." *Regional and Federal Studies* 9(1): 111–133.

King, George E. 2012. "Hydraulic Fracturing 101: What Every Representative, Environmentalist, Regulator, Reporter, Investor, University Researcher, Neighbor and Engineer Should Know about Estimating Frac Risk and Improving Frac Performance in Unconventional Gas and Oil Wells." *Society of Petroleum Engineers Conference Paper*. Woodlands, TX: Society of Petroleum Engineers. Paper No. 152596.

King, Pamela. 2015. "Eagle Ford Gas Extraction Results in Net Water Savings—Study." *Environment and Energy Publishing*, November 5. www.eenews.net/energywire/2015/11/05/stories/1060027472.

———. 2015a. "Methane Emissions Fall for 3rd Consecutive Year—EPA." *Environment and Energy Publishing*, October 7. www.eenews.net/energywire/2015/10/07/stories/1060025987.

Kingdon, John. 1995. *Agendas, Alternatives and Public Policies*. 2nd ed. New York: Harper Collins Publishers.

Kingston, Philip. 2014. "Kingston for Council." http://philipkingston.com/issues/.

Kleinhenz, Jack, and Russ Smith. 2011. "Ohio's Natural Gas and Crude Oil Exploration and Production Industry and the Emerging Utica Gas Formation." http://energyindepth.org/wp-content/uploads/ohio/2011/09/Ohio-Natural-Gas-and-Crude-Oil-Industry-Economic-Impact-Study-September-2011.pdf.

Klyza, Christopher, and David Sousa. 2007. *American Environmental Policy, 1990–2006*. Cambridge, MA: MIT Press.

Knox, Tom. 2015. "State's Biggest Oil and Gas Driller Cuts 15% of Its Ohio Work Force." *Columbus Business Journal*, September 30. www.bizjournals.com/columbus/blog/ohio-energy-inc/2015/09/states-biggest-oil-and-gas-driller-cuts-15-of-its.html.

———. 2014. "Ohio Finds Just 6 Cases of Water Problems from Drilling since 2010." *Columbus Business Journal*, January 7. www.bizjournals.com/columbus/news/2014/01/07/ohio-finds-just-6-cases-of-water.html?page=all.

Konisky, David M., and Neal D. Woods. 2012. "Measuring State Environmental Policy." *Review of Policy Research* 29(4): 544–569.

Konschnik, Katherine E., and Mark K. Boling. 2014. "Shale Gas Development: A Smart Regulation Framework." *Environmental Science & Technology* 48(15): 8404–8416.

Kooiman, Jan. 1993. *Modern Governance: New Government-Society Interactions*. ed. Jan Kooiman. New York: Sage Publications.

Kovac, Marc. 2013. "Kasich: GOP Fracking Tax Plan Not Enough." *Record Courier*, December 26. http://staging.recordpub.com/news%20local/2013/12/26/kasich-gop-fracking-tax-plan-not-enough (January 14, 2014).

Kraft, Michael, Mark Stephan, and Troy Abel. 2011. *Coming Clean: Information Disclosure and Environmental Performance*. Cambridge, MA: MIT University Press.

———, Platon N. Rigos, and Melvin B. Hill, Jr. (Eds.). 2001. *Home Rule in America: A Fifty-State Handbook*. Washington, DC: CQ Press.

Krauss, Clifford. 2015. "Low Oil Prices Pose Threat to Texas Fracking Bonanza." *New York Times*, August 14. www.nytimes.com/2015/08/15/business/energy-environment/low-oil-prices-pose-threat-to-texas-fracking-bonanza.html?_r=0.

Krueger, Skip, and Ethan M. Bernick. 2012. "State Rules and Local Governance Choices." *Publius: The Journal of Federalism* 40(4): 697–718.

Kuckro, Rod. 2016. "Ohio PUC Chairman Seeks Balance in Tackling Carbon Emissions." *Environment and Energy Publishing*, June 14. www.eenews.net/stories/1060038751.

———. 2013. "EIA Forecast Reflects Seismic Changes in Electricity Demand." *Environment and Energy Publishing*, December 17. www.eenews.net/stories/1059991960.

Kuster, Kent. 2012. *Summary of Colorado Public Health Studies 2005–2012*. Colorado Department of Public Health and Environment. www.colorado.gov/pacific/sites/default/files/OGC_Health-Studies.pdf (February 3, 2014).

Ladd, Anthony E. 2013. "Stakeholder Perceptions of Socio-Environmental Impacts from Unconventional Natural Gas Development and Hydraulic Fracturing in the Haynesville Shale." *Journal of Rural Social Sciences* 28(2): 56–89.

Lavelle, Marianne. 2015. "5 Ways Energy Is Transforming U.S. Railroads: Trains Aren't Just Carrying Crude Oil When It Comes to Energy." *Scientific American*, March 16. www.scientificamerican.com/article/5-ways-energy-is-transforming-u-s-railroads/.

Layzer, Judith A. 2002. *The Environmental Case: Translating Values into Policy*. Washington, DC: CQ Press.

Lehner, Peter H. 1993. "Act Locally: Municipal Enforcement of Environmental Law." *Stanford Environmental Law Journal* 12: 50–81.

Leighninger, Matthew. 2006. *The Next Form of Democracy: How Expert Rule Is Giving Way to Shared Governance—and Why Politics Will Never be the Same*. Nashville, TN: Vanderbilt University Press.

Lemos, Maria Carmen, and Arun Agrawal. 2006. "Environmental Governance." *Annual Review of Environment and Resources* 31: 297–325.

Lewandowski, Brian, and Richard Wobbekind. 2013. "Assessment of Oil and Gas Industry 2012: Industry Economic and Fiscal Contributions in Colorado." www.coga.org/wp-content/uploads/2015/09/2012-Economic-Study-CU-Leeds-School-of-Business.pdf.

Liesik, Geoff. 2015. "Ute Tribe seeks to Join Lawsuit against BLM 'Fracking' Rule." *Deseret News*, June 23. www.deseretnews.com/article/865631271/Ute-Tribe-seeks-to-join-lawsuit-against-BLM-fracking-rule.html?pg=all.

Lisik, Brian. 2013. "Brunswick Council Passes Resolution Addressing Fracking." *Sun News*, February 26. www.neogap.org/neogap/2013/02/26/brunswick-council-passes-resolution-addressing-fracking/.

Lofholm, Nancy. 2015. "Fracking Paused on Gas Wells in Western Colorado." *The Denver Post*, January 23. www.denverpost.com/business/ci_27382264/fracking-paused-gas-wells-western-colorado.

Loftis, Randy. 2013. "Dallas OKs Gas Drilling Rules That Are among Nation's Tightest" *Dallas Morning News*, December 12. www.dallasnews.com/news/metro/20131211-dallas-new-gas-drilling-rules-are-among-nations-tightest.ece (December 29, 2013).

Long, Norton. 1949. "Power and Administration." *Public Administration Review* 9(4): 257–264.

Louis, Brian. 2012. "Fracking in Ohio Sparks Real Estate Rebound: Mortgages." *Bloomberg News*, June 11. www.bloomberg.com/news/2012-06-11/fracking-in-ohio-sparks-real-estate-rebound-mortgages.html.

Lovell, T. Lynn, Joe Barrow, and William A. Wiegand. 2008. "Permitting of Natural Gas Wells in the Floodplain." *World Environmental and Water Resources Congress 2008*: 1–10.

Lovrich, Nicholas P., and Meredith A. Newman. 2004. "The Hearing of Local Government Interests in State Legislatures: The Effects of Prior Service in City or County Government." *State & Local Government Review* 36(1): 67–77.

Lowry, William. 1991. *The Dimensions of Federalism: State Governments and Pollution Control Policies*. Durham, NC: Duke University Press.

Lubell, Mark. 2002. "Environmental Activism as Collective Action." *Environment and Behavior* 34(4): 431–454.

Lynn, Steve. 2012. "Cities Take Aim at Oil Industry Regulations." *Northern Colorado Business Review*, June 29. www.ncbr.com/article/20120629/EDITION/120629913/-1/oilnaturalgas.

McAdam, Doug, and Hilary S. Boudet. 2012. *Putting Social Movements in Their Place*. New York: Cambridge University Press.

McCright, Aaron M., and Riley Dunlap. 2011. "The Politicization of Climate Change and Polarization in the American Public's Views of Global Warming." *The Sociological Quarterly* 52(1): 155–194.

McFarland, John B. 2013. "Role of Localities and the Public in Shaping Drilling's Future: A View from Texas (The American Law Institute Continuing Legal Education Series)." http://files.alicle.org/thumbs/datastorage/skoobesruoc/pdf/TSUX04_chapter_04_thumb.pdf.

McGuire, Michael. 2006. "Intergovernmental Management: A View from the Bottom." *Public Administration Review* 66(5): 677–679.

McKenzie, Lisa M., Roxana Z. Witter, Lee S. Newman, and John L. Adgate. 2012. "Human Health Risk Assessment of Air Emissions from Development of Unconventional Natural Gas Resources." *Science of the Total Environment* 424(May): 79–87.

McKinney, Matthew, and Will Harmon. 2002. "Public Participation in Environmental Decision Making: Is It Working?" *National Civic Review* 91(2): 149–170.

McParland, Tom. 2014. "Policy Matters Ohio, Hagan Take Aim at GOP Severance Tax Bill." *The Vindicator*, January 15. www.vindy.com/news/2014/jan/15/report-blasts-gop-severance-tax-bill/#sthash.p3F5LflP.dpuf.

———. 2014a. "ODNR Officials Preview New Oil and Gas Regs." *The Vindicator*, January 17. www.vindy.com/news/2014/jan/17/regional-chamber-luncheon-officials-prev/.

Maher, Jessica. 2014. "Voters Reject Loveland Fracking Moratorium." *Loveland Reporter-Herald*, June 24. www.reporterherald.com/news/loveland-local-news/ci_26028522/voters-reject-loveland-fracking-moratorium.

———. 2013. "Loveland City Council Passes Oil and Gas Regulations on Second Reading: New Regulations Will Take Effect April 2." *Loveland Reporter-Herald*, March 20. www.reporterherald.com/news/loveland-local-news/ci_22828085/loveland-city-council-passes-oil-and-gas-regulations?source=most_viewed .

Main, Douglas. 2013. "Fracking Wastewater Contaminated—and Likely Radioactive." *NBC News*, October 3. www.nbcnews.com/science/fracking-wastewater-contaminated-likely-radioactive-8C11323012.

Makarewicz, Julie. 2012. "Thornapple Township Officials Urge Ban on Fracking." *The Hastings Banner*, June 21. http://hastingsbanner.com/thornapple-township-officials-urge-ban-on-fracking-p3875-87.htm#sthash.dv9iDWYb.dpuf.

Malewitz, Jim. 2015. "Texas House Approves 'Denton Fracking Bill'." *The Texas Tribune*, April 17. www.wfaa.com/story/news/politics/2015/04/17/texas-house-to-drill-into-denton-fracking-bill/25926191/.

———. 2014. "Denton Group Seeking Fracking Ban Cites Gains." *The Texas Tribune*, March 14. www.texastribune.org/2014/03/14/denton-group-seeks-local-fracking-ban/.

———. 2013. "Dallas City Council Tightens Gas Drilling Ordinance." *The Texas Tribune*, December 11. www.texastribune.org/2013/12/11/dallas-city-council-tightens-gas-drilling-ordinanc/.

———. 2013a. "Dallas Debate Shows City Power to Limit Drilling." *The Texas Tribune*, October 18. www.texastribune.org/2013/10/18/dallas-debate-shows-city-power-limit-drilling/.

Marmaduke, Jacy. 2016. "High Court Strikes Down Fort Collins' Halt to Fracking." *Coloradoan*, May 4. www.coloradoan.com/story/news/2016/05/02/colorado-supreme-court-rules-against-fort-collins-fracking-moratorium/83798238/.

Marshall, Aaron. 2012. "Mansfield Leads Legal Fight Against Injection Wells Holding Fracking Waste." *Cleveland Plain-Dealer*, September 10. www.cleveland.com/open/index.ssf/2012/09/mansfield_leads_legal_fight_ag.html.

———. 2012a. "New Wave of Injection Wells on the Way in Ohio for Fracking Waste." *Cleveland Plain-Dealer*, November 23. www.cleveland.com/open/index.ssf/2012/11/new_wave_of_injection_wells_on.html.

Mashaw, Jerry. 2006. "Accountability and Institutional Design: Some Thoughts on the Grammar of Governance." In *Public Accountability: Designs, Dilemmas and Experiences*. ed. Michael Dowdle. Cambridge, MA: Cambridge University Press.

Matisoff, Daniel C. 2008. "The Adoption of State Climate Change Policies and Renewable Portfolio Standards: Regional Diffusion or Internal Determinants?" *Review of Policy Research* 25(6): 527–546.

Matland, Richard. 1995. "Synthesizing the Implementation Literature: The Ambiguity-Conflict Model of Policy Implementation." *Journal of Public Administration Research and Theory* 5(2): 145–174.

Maxwell, Billie Ann. 2009. "Texas Tug of War: A Survey of Urban Drilling and the Issues an Operator Will Face." *Texas Journal of Oil, Gas, and Energy Law* 4(2): 337–363.

May, Peter J. 1992. "Policy Learning and Failure." *Journal of Public Policy* 12(4): 331–354.

Miller, Chelsea. 2013. "Oberlin Fracking Ban to Appear on Ballot." *The Chronicle-Telegram*, August 7. http://chronicle.northcoastnow.com/2013/08/07/oberlin-fracking-ban-to-appear-on-ballot/.

Miller, Craig. 2015. "Interior Secretary: Local Fracking Bans Are 'Wrong Way to Go'." *KQED*, January 2. http://ww2.kqed.org/science/2015/01/02/interior-secretary-local-fracking-bans-are-wrong-way-to-go/.

Miller, Joanne M., and Jon A. Krosnick. 2004. "Threat as a Motivator of Political Activism: A Field Experiment." *Political Psychology* 25(4): 507–523.

Minor, Joel. 2014. "Local Government Fracking Regulations: A Colorado Case Study." *Stanford Environmental Law Journal* 33: 61–122.

Moe, Terry. 1989. "The Politics of Bureaucratic Structure." In *Can the Government Govern?* eds. John E. Chubb and Paul E. Peterson. Washington, DC: Brookings Institution Press.

Mosier, Samantha, and Jonathan M. Fisk. 2013. "Can Local Voluntary Environmental Programs 'Work'? An Examination of Fort Collins' Climate Wise Program." *Environmental Management* 51(5): 969–987.

Mosqueda, Priscila. 2013. "Dallas City Council Rejects Fracking." *Texas Observer*, August 28. www.texasobserver.org/dallas-city-council-rejects-fracking/

Mouawad, Jad. 2015. "New Oil Train Rules Are Hit from All Sides." *New York Times*, May 1. www.nytimes.com/2015/05/02/business/energy-environment/us-sets-new-rules-for-oil-shipments-by-rail.html?_r=0.

Muehlenbachs, Lucija, Elisheba Spiller, and Christopher Timmins. 2015. "The Housing Market Impacts of Shale Gas Development." *The American Economic Review* 105(12): 3633–3659.

———. 2013. "Shale Gas Development and the Costs of Groundwater Contamination Risk." *Resources for the Future Discussion Paper, RFF DP 12–40-REV 12–40*. Cambridge, MA: National Bureau of Economic Research. www.rff.org/files/sharepoint/WorkImages/Download/RFF-DP-12-40-REV.pdf.

Mueller, Laura. 2015. "Alphabet Soup: Types of Cities." www.tml.org/p/typesofcities.pdf.

Mufson, Steven. 2013. "Obama Administration Issues Draft Fracking Regulations." *Washington Post*, May 16. http://articles.washingtonpost.com/2013-05-16/business/39305945_1_fracking-regulations-fracfocus-hydraulic-fracturing

National Association of State Budgetary Officers. 2015. "State Severance Tax Revenue and the Falling Price of Oil." www.nasbo.org/budget-blog/state-severance-tax-revenue-and-falling-price-oil.

National Conference of State Legislatures [NCSL]. 2014. "Transporting Crude Oil by Rail: State and Federal Action." www.ncsl.org/research/energy/transporting-crude-oil-by-rail-state-and-federal-action.aspx.

———. 2010. "Natural Gas: Smarter Power Today." www.ncsl.org/documents/energy/TommyFoltz.pdf.

National League of Cities [NLC]. 2013. "Local Government Authority." www.nlc.org/build-skills-and-networks/resources/cities-101/city-powers/local-government-authority.

National Public Radio [NPR]. 2013. "What the New Impact Fee Law, Act 13, Means for Pennsylvania." http://stateimpact.npr.org/pennsylvania/tag/impact-fee/

NBC-DFW. 2014. "Denton Pushes Back in Court Over Fracking Lawsuits." *NBC*, December 1, www.nbcdfw.com/news/local/Denton-Pushes-Back-in-Court-Over-Fracking-Lawsuits-284405161.html.

Negro, Sorrell. 2012. "Fracking Wars: Federal, State and Local Conflicts over the Regulation of Natural Gas Activities." *Zoning and Planning Law Report* 35(2): 1–13.

Nerurkar, Neelesh. 2012. "U.S. Oil Imports and Exports." www.fas.org/sgp/crs/misc/R42465.pdf.

Neuman, Scott. 2015. "Interior Department Issues New Fracking Rules For Federal Lands." *National Public Radio*, March 20. www.npr.org/sections/thetwo-way/

2015/03/20/394282086/interior-dept-issues-new-fracking-rules-for-federal-lands.
News 4. 2015. "Texas Steps up Monitoring for Fracking Air Pollution." *News 4*. http://news4sanantonio.com/m/news/features/trouble-shooters/stories/Texas-steps-up-monitoring-for-fracking-air-pollution-84090.shtml.
Nichols, Chris. 2015. "Texas Prepares for Drop in Oil-and-Gas Tax Revenue." *Yahoo Finance*, January 12. http://finance.yahoo.com/news/texas-prepares-for-drop-in-oil-and-gas-tax-revenue-210205520.html.
Nicot, Jean-Philippe, and Bridget R. Scanlon. 2012. "Water Use for Shale-Gas Production in Texas, US." *Environmental Science & Technology* 46(6): 3580–3586.
Niquette, Mark. 2012. "Ohio City Looks to Fracking as Way to Raise Money for Demolition." *Bloomberg News*, September 26. www.bloomberg.com/news/2012-09-27/ohio-city-looks-to-fracking-as-way-to-raise-money-for-demolition.html/
———. 2012a. "Ohio Mayor Buys Quake Insurance as He Seeks Answers on Fracking." *Bloomberg News*, January 4. www.bloomberg.com/news/2012-01-04/ohio-mayor-buys-quake-insurance-as-he-seeks-answers-on-fracking.html.
Nolon, John R., and Steven E. Gavin. 2013. "Hydrofracking: State Preemption, Local Power and Cooperative Governance." Case Western Reserve Law Review 63: 995-1036.
Noon, Marita. 2013. "Why Punish Oil and Gas Producers." *Albuquerque Journal*, September 2. www.abqjournal.com/256533/opinion/why-punish-oil-and-gas-producers.html.
Obama, Barack. 2013. "Remarks by the President in the State of the Union Address." www.whitehouse.gov/the-press-office/2013/02/12/remarks-president-state-union-address.
Observer Staff. 2013. "Analysis: In Local Fracking Wars, Tide Turns against Green Lobby." *The Colorado Observer*, June 19. http://thecoloradoobserver.com/2013/06/analysis-green-lobby-meets-local-resistance-in-fracking-fights/.
Occupational Safety and Health Administration [OSHA]. 2015. "OSHA Hazard Alert." www.osha.gov/dts/hazardalerts/hydraulic_frac_hazard_alert.html.
———. 2015a. "Oil and Gas Drilling." www.osha.gov/SLTC/oilgaswelldrilling/.
Office of Conservation, Department of Natural Resources. 2014. "Regulation of Oil and Gas Wells and Management of Orphaned Wells." http://app.lla.state.la.us/PublicReports.nsf/0/D6A0EBE279B83B9F86257CE700506EAD/$FILE/000010BC.pdf.
Ogburn, Stephanie P. 2014. "Colorado First State to Limit Methane Pollution from Oil and Gas Wells: Air Pollution from Fossil Fuel Production Has Contributed to Worsening Smog and Climate Change." *Scientific American*, February 25. www.scientificamerican.com/article/colorado-first-state-to-limit-methane-pollution-from-oil-and-gas-wells/.
Ohio Department of Job and Family Services. 2016. "Quarterly Economic Trends for Ohio Oil and Gas Industries." http://ohiolmi.com/OhioShale/Ohio%20Shale%20Report_2Q_2015.pdf.
Ohio Department of Natural Resources [ODNR]. 2016. "Laws and Regulations Summary." http://oilandgas.ohiodnr.gov/laws-regulations/oil-gas-law-summary.

———. 2016a. "Oil and Gas Well Viewer." https://gis.ohiodnr.gov/website/dog/oilgaswells/#.

———. 2015. "Ice, Dust Control, and Brine." http://oilandgas.ohiodnr.gov/local-governments/ice-dust-control-brine.

———. 2014. "Fact Sheet." http://oilandgas.ohiodnr.gov/citizens/oil-gas-faq.

Ohio Department of Taxation. N.d. "Severance Tax." www.tax.ohio.gov/Portals/0/communications/publications/annual_reports/2014_annual_report/2014_AR_Section_2_Severance_Tax.pdf.

Ohio Environmental Council. 2014. "Fracking Impacts and Air Quality." www.theoec.org/campaign/fracking-impacts-air-quality.

Ohio Environmental Protection Agency [OEPA]. 2014. "About Us." www.epa.state.oh.us/about.aspx.

———. 2014a. "Marcellus and Utica Shale." www.epa.state.oh.us/MarcellusandUticaShale.aspx.

———. 2014b. "Rules and Laws." www.epa.ohio.gov/rules.aspx.

Ohio Oil and Gas Association. 2015. "2015 Oil & Gas Activity in Ohio." http://c.ymcdn.com/sites/www.ooga.org/resource/resmgr/Files/The_Debrosse_Memorial_Report.pdf.

———. 2013. "Overview of Ohio Oil & Natural Gas." www.aeeohio.com/OOGA_SHALE%20OIL%20AND%20GAS_05022103.pdf.

———. 2012. "Ohio Shale Plays." www.ooga.org/?page=OhioShalePlays.

Oldham, Jennifer. 2014. "Colorado First State to Clamp Down on Fracking Methane Pollution." *Bloomberg News*, February 23. www.bloomberg.com/news/articles/2014-02-24/colorado-first-state-to-clamp-down-on-fracking-methane-pollution.

Opp, Susan M. 2012. "Environmental Justice and the Resource Conservation Recovery Act Inspection and Enforcement Process." *International Review of Public Administration* 17(1): 179–192.

———, and Kyle L. Saunders. 2013. "Pillar Talk: Local Sustainability Initiatives and Policies in the United States—Finding Evidence of the Three E's: Economic Development, Environmental Protection, and Social Equity." *Urban Affairs Review* 49(5): 678–717.

Oppel, Richard A., Jr. 2015. "Oklahoma Court Rules Homeowners Can Sue Oil Companies Over Quakes." *New York Times*, June 30. www.nytimes.com/2015/07/01/us/oklahoma-court-rules-homeowners-can-sue-oil-companies-over-quakes.html?_r=0.

Opsal, Tara, and Tara O'Connor Shelley. 2014. "Energy Crime, Harm, and Problematic State Response in Colorado: A Case of the Fox Guarding the Hen House?" *Critical Criminology* 22(4): 561–577.

Ostrom, Elinor. 2010. "A Long Polycentric Journey." *Annual Review of Political Science* 13: 1–23.

———. 2009. "A Polycentric Approach for Coping with Climate Change." *World Bank Policy Research Working Paper Series*. http://ssrn.com/abstract=1494833.

———. 1990. *Governing the Commons: The Evolution of Institutions for Collective Action*. Cambridge, MA: Cambridge University Press.

O'Toole, Laurence J., Jr. 1997. "Treating Networks Seriously: Practical and Research-Based Agendas in Public Administration." *Public Administration Review* 57(1): 45–52.

Ouzts, Elizabeth. 2015. "Ohio Local Officials Want to Control Fracking." www.environmentamerica.org/news/ame/ohio-local-officials-want-control-fracking.

Owens, Katharine, and Carl Zimmerman. 2013. "Local Governance Versus Centralization: Connecticut Wetlands Governance as a Model." *Review of Policy Research* 30(6): 629–656.

Pace, Jessica. 2016. "Houston Energy Company to Close Durango Office." *Durango Herald*, January 25. www.durangoherald.com/article/20160125/NEWS01/160129727/Houston-energy-company-to-close-Durango-office.

Parfomak, Paul W. 2013. "Keeping America's Pipelines Safe and Secure: Key Issues for Congress." www.fas.org/sgp/crs/homesec/R41536.pdf.

Paulik, L. Blair, Carey E. Donald, Brian W. Smith, Lane G. Tidwell, Kevin A. Hobbie, Laurel Kincl, Erin N. Haynes, and Kim A. Anderson. 2016. "Emissions of Polycyclic Aromatic Hydrocarbons from Natural Gas Extraction into Air." *Environmental Science & Technology* 50(14): 7921–7929.

Pelzer, Jeremy. 2015. "Lawmakers Say Ohio's Oil and Gas Tax Is among Nation's Lowest, but Urge Caution on Increase." *Cleveland Plain-Dealer*, October 23. www.cleveland.com/open/index.ssf/2015/10/lawmakers_say_ohios_oil_and_ga.html.

———. 2014. "Severance Tax Compromise Could be Close, but Ohio Lawmakers Might Hold Off on Tax Proposals Until Fall." *Cleveland Plain Dealer*, March 25. www.cleveland.com/open/index.ssf/2014/03/severance_tax_compromise_could.html.

Perryman Group. 2014. "The Economic and Fiscal Contribution of the Barnett Shale: Impact of Oil and Gas Exploration and Production on Business Activity and Tax Receipts in the Region and State." www.fortworthchamber.com/chamber/docs/BSISeptember2014.pdf.

Personal Communication. Local Official 1, 3/1/2016.

Personal Communication. Local Official 2, 4/25/2016.

Personal Communication. State Official 1, 6/6/2014.

Peters, B. Guy. 2011. *Institutional Theory in Political Science: The New Institutionalism*. New York: Bloomsbury Publishing USA.

Peterson, Matt. 2015. "Mansfield Council Approves New Gas Drilling Ordinance That Won't Widen Buffer." *Dallas Morning News*, March 24. http://thescoopblog.dallasnews.com/2015/03/mansfield-council-approves-new-ordinance-on-gas-drilling-but-wont-widen-buffer.html/.

Peterson, Paul E. 1995. *The Price of Federalism*. Washington, DC: Brookings Institution Press.

Phillips, Andrew, Robert Cline, Caroline Sallee, Michelle Klassen, and Daniel Sufranski. 2013. "Total State and Local Business Taxes State-by-State Estimates for Fiscal Year 2012." www.cost.org/workarea/downloadasset.aspx?id=84767.

Phillips, Susan. 2015. "Preparing for the Worst, Delco First Responders Simulate Oil Train Accident." *National Public Radio*, March 24. https://stateimpact.npr.org/pennsylvania/2015/03/24/preparing-for-the-worst-delco-first-responders-simulate-oil-train-accident/.

———. 2014. "Maryland Governor Proposes Stringent Fracking Regulations." *National Public Radio*, November 11. http://stateimpact.npr.org/pennsylvania/2014/11/25/maryland-governor-proposes-stringent-fracking-regulations/.

———. 2013. "Pennsylvania Communities Craft Creative Escape Hatch from Drilling Law." http://stateimpact.npr.org/pennsylvania/2013/02/13/pa-communities-craft-creative escape-hatch-from-drilling-law/.

———. 2012. "Four Townships Say They're Still Owed Impact Fee Money." http://stateimpact.npr.org/pennsylvania/2012/10/19/four-townships-say-theyre-still-owed-impact-fee-money/.

Pipeline and Hazardous Materials Safety Administration [PHMSA]. 2015. "Dashboard." https://hip.phmsa.dot.gov/analyticsSOAP/saw.dll?Dashboard.

———. 2014. "Operation Safe Delivery Report." www.phmsa.dot.gov/pv_obj_cache/pv_obj_id_8A422ABDC16B72E5F166FE34048CCCBFED3B0500/filename/07_23_14_Operation_Safe_Delivery_Report_final_clean.pdf.

Pless, Jacqueline. 2012. "Fracking Fracas." www.ncsl.org/research/energy/fracking-fracas.aspx.

———. 2012a. "Oil and Gas Severance Taxes: States Work to Alleviate Fiscal Pressures Amid The Natural Gas Boom." www.ncsl.org/research/energy/oil-and-gas-severance-taxes.aspx.

———. 2011. "Fracking Update: What States Are Doing to Ensure Safe Natural Gas Extraction." www.ncsl.org/research/energy/fracking-update-what-states-are-doing.aspx.

Portney, Kent. 2009. "Sustainability in American Cities: A Comprehensive Look at What Cities are Doing and Why?" In *Toward Sustainable Communities*. eds. Daniel A. Mazmanian and Michael E. Kraft. Cambridge, MA: MIT Press.

———. 2003. *Taking Sustainable Cities Seriously*. Cambridge, MA: MIT Press.

———, and Jeffrey M. Berry. 2010. "Participation and the Pursuit of Sustainability in U.S. Cities." *Urban Affairs Review* 46(1): 119–139.

Prah, Pamela. 2013. "Fracking for State Dollars." *Stateline: The Daily News Service of the Pew Charitable Trusts*, January 31. www.pewstates.org/projects/stateline/headlines/fracking-for-state-dollars-85899447047 (October 14, 2013).

Pralle, Sarah. 2006. *Branching Out, Digging In*. Washington, DC: Georgetown University Press.

Pressman, Jeffrey, and Aaron Wildavsky. 1984. *Implementation: How Great Expectations in Washington are Dashed in Oakland; or, Why It's Amazing That Federal Programs Work at All, This Being a Saga of the Economic Development Administration as Told by Two Sympathetic Observers Who Seek to Build Morals on a Foundation of Ruined Hopes*. Berkeley, CA: University of California Press.

Price, Asher. 2012. "Railroad Commissioner David Porter Helping to Shape Fracking Proposals amid Boom." *Austin Statesman*, September 22. www.statesman.com/news/news/state-regional/railroad-commissioner-david-porter-helping-to-sh-1/nSLbt/

Proctor, Cathy. 2013. "Encana Will Halt Western Colorado Drilling in 2014." *Denver Business Journal*, December 16. www.bizjournals.com/denver/blog/earth_to_power/2013/12/encana-calls-halt-to-drilling-in.html?page=all.

———. 2012. "Energy Industry a Force of Nature across Colorado." *Denver Business Journal*, November 30. www.bizjournals.com/dallas/print-edition/2012/11/30/Colorado-energy-industry-a-force.html?page=all.

Putnam, Robert D. 2000. *Bowling Alone: The Collapse and Revival of American Community*. New York: Simon and Schuster.

Quinn, Megan. 2013. "Fracking Takes Center Stage Wednesday: Large Crowd Expected for Town Hall Forum at Auditorium." *Broomfield Enterprise*,

May 18. www.broomfieldenterprise.com/ci_23268109/fracking-takes-center-stage-wednesday-at-broomfield-auditorium.
Rabe, Barry G. 2010. "Racing to the Top, the Bottom, or the Middle of the Pack? Evolving State Gov't Role in Environmental Protection." In *Environmental Politics & Policy: New Directions for the 21st Century*. 8th ed. eds. Norman Vig and Michael Kraft. Washington, DC: CQ Press.
———. 2006. *Second Generation Climate Policies in the American States: Proliferation, Diffusion and Regionalization*. Washington, DC: The Brookings Institution.
———. 2004. *Statehouse & Greenhouse: The Emerging Politics of American Climate Change Policy*. Washington, DC: Brookings Institution Press.
———, and Christopher Borick. 2013. "Conventional Politics for Unconventional Drilling? Lessons from Pennsylvania's Early Move into Fracking Policy Development." *Review of Policy Research* 30(3): 321–340.
———, and Rachel Hampton. 2015. "Taxing Fracking: The Politics of State Severance Taxes in the Shale Era." *Review of Policy Research* 32(4): 389–412.
Rahm, Diane. 2011. "Regulating Hydraulic Fracturing in Shale Gas Plays: The Case of Texas." *Energy Policy* 39(5): 2974–2981.
Railroad Commission of Texas. 2016. "Historical Production Data." www.rrc.state.tx.us/oil-gas/research-and-statistics/production-data/historical-production-data/natural-gas-production-and-well-counts-since-1935/.
———. 2016a. "Oil and Gas Enforcement Data." www.rrc.state.tx.us/media/24258/fy-14-oil-and-gas-enforcement-data-revised-4th-qtr.pdf.
———. 2015. "News." www.rrc.state.tx.us/all-news/100515a/.
———. 2014. "Frequently Asked Questions." www.rrc.state.tx.us/about/faqs/hydraulicfracturing.php.
———. 2014a. "Injection and Disposal Wells." www.rrc.state.tx.us/about-us/resource-center/faqs/oil-gas-faqs/faq-injection-and-disposal-wells/.
———. 2014b. "Spills." www.rrc.state.tx.us/environmental/spills/h8s/.
———. 2014c. "Texas Monthly Oil and Gas Statistics." www.rrc.state.tx.us/all-news/062514a/.
Raimi, Daniel, and Richard G. Newell. 2014. "Shale Public Finance." http://energy.duke.edu/sites/energy.duke.edu/files/files/Shale%20Public%20Finance%20Local%20Revenues%20and%20Costs%20Report%20Final.pdf.
Ramsey, Ross, and Neena Satija. 2013. "Texas Voters Approve Nine Constitutional Amendments." *The Texas Tribune*, November 6. www.texastribune.org/2013/11/06/election-results-headline/.
Randall, Bob. 2012. "Colorado Oil and Gas Conservation Commission Rules and Regulations: BLM Colorado Resource Advisory Councils." www.blm.gov/pgdata/etc/medialib/blm/co/resources/resource_advisory/2012_super_rac.Par.83149.File.dat/COGCC.pdf.
Rawlins, Rachael A. 2014. "Planning for Fracking on the Barnett Shale: Urban Air Pollution, Improving Health Based Regulation, and the Role of Local Governments." *Virginia Environmental Law Journal* (31): 227–306.
Redd, G. 2013. "Illinois Governor Signs Strictest Fracking Law in Nation." *The Jurist*, June 19. http://jurist.org/paperchase/2013/06/illinois-governor-signs-strictest-fracking-law-in-nation.php.

Redden, Molly. 2013. "Scott Walker's Sand Grab: Wisconsin Wants a Piece of the Fracking Boom, No Matter Who Gets Hurt." *New Republic*, August 21. www.newrepublic.com/article/114320/frac-sand-mining-wisconsin-rides-fracking-boom.

Richardson, Jesse J., Jr. 2011. "Dillon's Rule Is from Mars, Home Rule Is from Venus: Local Government Autonomy and the Rules of Statutory Construction." *Publius: The Journal of Federalism* 41(4): 662–685.

Richardson, Nathan, Madeline Golieb, Alan Krupnick, and Hannah Wiseman. 2013. "The State of State Shale Gas Regulation: State-by-State Tables." www.rff.org/files/document/file/RFF-Rpt-StateofStateRegs_ExecSumm_0.pdf.

Ridlington, Elizabeth, and John Rumpler. 2013. "Fracking by the Numbers: Key Impacts of Dirty Drilling at the State and National Level." www.environmentamerica.org/sites/environment/files/reports/EA_FrackingNumbers_scrn.pdf.

Riley, Timothy. 2007. "Wrangling with Urban Wildcatters: Defending Texas Municipal Oil and Gas Development Ordinances against Regulatory Takings Challenges." *Vermont Law Review* 32: 349–407.

Ringquist, Evan. 1993. *Environmental Protection at the State Level: Politics and Progress in Controlling Pollution*. Armonk, NY: M.E. Sharpe Inc.

———, and James Garand. 1999. "Policy Change in the American States." In *American State and Local Politics: Directions for the 21st Century*. eds. Ronald Weber and Paul Brace. Chatham, NJ: Chatham House.

Riverstone-Newell, Lori. 2012. "Bottom-Up Activism: A Local Political Strategy for Higher Policy Change." *Publius: The Journal of Federalism* 42(3): 401–421.

Roberts, Paul. 2004. *The End of Oil: On the Edge of a Perilous New World*. New York: Mariner Books.

Rochat, Scott. 2014. "Boulder County Judge Strikes Down Longmont Fracking Ban: Order Stayed Pending Possible Appeal." *The Denver Post*, July 24. www.denverpost.com/news/ci_26209898/boulder-county-judge-strikes-down-longmont-fracking-ban.

———. 2013. "Colorado's Oil/Gas Lawsuit against Longmont Set for August Trial First and Main Zoning Approved Again." *Longmont Times-Call*, October 28. www.timescall.com/longmont-local-news/ci_24406905/colorados-oil-gas-lawsuit-against-longmont-set-august

———. 2012. "Longmont Council Approves Oil/Gas Rules 5–2." *Longmont Times-Call*, July 17. www.timescall.com/news/longmont-local-news/ci_21098770/longmont-council-approves-oil-gas-rules-5-2.

———. 2012a. "Longmont Voters Pass Ban on Fracking." *The Denver Post*, November 8. www.denverpost.com/politics/ci_21952523/longmont-voters-pass-ban-fracking?.

———. 2012b. "Spending Passes $500,000 to Fight Fracking Ban." Boulder Daily Camera, October 24. www.dailycamera.com/news/ci_21847759/spending-passes-500-000-fight-fracking-ban

Rochester Business Journal. 2014. "Most Readers Disagree with Fracking Ban." *Rochester Business Journal*, December 26. www.rbj.net/article.asp?aID=212555.

Rogers, Paul. 2015. "Fracking: Oil Company Drops Lawsuit Attempting to Overturn San Benito County Ban." *The Mercury News*, April 6. www.mercurynews.com/2015/04/06/fracking-oil-company-drops-lawsuit-attempting-to-overturn-san-benito-county-ban/.

Rozboril, Robert. 2012. "Broadview Heights Issue 29 Passes, Mayor Fears It Will Create More Problems for the City." *Cleveland Plain-Dealer*, November 7. www.cleveland.com/broadview-heights/index.ssf/2012/11/broadview_heights_issue_29_pas.html.

Ruddell, Rick, Dheeshana S. Jayasundara, Roni Mayzer, and Thomasine Heitkamp. 2014. "Drilling Down: An Examination of the Boom-Crime Relationship in Resource Based Boom Counties." *Western Criminology Review* 15(1): 3–17.

Rutherford, Jake. 2014. "Don't Frac This Up: Denton's Frac Ban and the Appropriate State Legislative Response." *Texas Tech Law Review* 47: 844–890.

Sabatier, Paul. 1986. "Top-Down and Bottom-Up Approaches to Implementation Research: A Critical Analysis and Suggested Synthesis." *Journal of Public Policy* 6(1): 21–48.

Saha, Devashree, and Robert G. Paterson. 2008. "Local Government Efforts to Promote the 'Three E's' of Sustainable Development: Survey in Medium to Large Cities in the United States." *Journal of Planning Education and Research* 28(21): 21–37.

Sakelaris, Nicholas. 2015. "Bill That Would Ban Fracking Bans Breezes through the Texas House." *Dallas Business Journal*, April 17. www.bizjournals.com/dallas/news/2015/04/17/bill-that-would-ban-fracking-bans-breezes-through.html.

Salazar, Daniel. 2014. "As Fracking Looms, North Carolina Officials Worry about Road Damage." *McClatchy DC*, November 18. www.mcclatchydc.com/2014/11/18/247261/as-fracking-looms-north-carolina.html.

Salley, Mark. 2016. "New Data to be Used to Assess Health Risks from Oil and Gas Operations." www.colorado.gov/pacific/cdphe/news/oil-and-gas-data.

Sandrick, Bob. 2015. "Judge Shoots Down Broadview Heights Ban on Future Oil and Gas Wells." *Cleveland Plain-Dealer*, March 16. www.cleveland.com/broadview-heights/index.ssf/2015/03/judge_shoots_down_broadview_he.html.

———. 2015a. "Ohio Oil and Gas Commission Revokes Drilling Permit for North Royalton Well." *Cleveland Plain-Dealer*, May 8. www.cleveland.com/north-royalton/index.ssf/2014/12/ohio_oil_and_gas_commission_re.html/.

———. 2013. "North Royalton Joins Broadview Heights in Home-Rule, Fracking Battle in Ohio Supreme Court." *Cleveland Plain-Dealer*, September 14. www.cleveland.com/north-royalton/index.ssf/2013/09/north_royalton_joins_broadview.html.

Satija, Neena, and Becca Aaronson. 2014. "Visualization: Water for Fracking in Ten Texas Counties." *The Texas Tribune*, February 18. www.texastribune.org/2014/02/18/water-fracking-counties/.

Scheberle, Denise. 2004. *Federalism and Environmental Policy*. 2nd ed. Washington, DC: Georgetown University Press.

Schlosberg, David. 2007. *Defining Environmental Justice*. Oxford: Oxford University Press.

Schmall, Emily. 2014. "Texas Gas Town Considers Ban on Fracking." *The Durango Herald*, June 20. www.durangoherald.com/article/20140622/NEWS03/140629992/Texas-gas-town-considers-ban-on-fracking-.

Schneider, Saundra K., William G. Jacoby, and Daniel C. Lewis. 2011. "Public Opinion toward Intergovernmental Policy Responsibilities." *Publius: The Journal of Federalism* 41(1): 1–30.

Scholz, John, and Cheng-Lung Wang. 2006. "Cooptation or Transformation? Local Policy Networks and Federal Regulatory Enforcement." *American Journal of Political Science* 50(1): 81–97.
Selman, Paul. 2004. "Community Participation in the Planning and Management of Cultural Landscapes." *Journal of Environmental Planning and Management* 47(3): 365–392.
Seville, Lisa R. 2014. "Danger Beneath: 'Fracking' Gas, Oil Pipes Threaten Rural Residents." *NBC News*, August 25. www.nbcnews.com/news/investigations/danger-beneath-fracking-gas-oil-pipes-threaten-rural-residents-n187021.
Shale Info. 2014. "Summary of ODNR & Ohio EPA Regulatory Authority Over Oil/Gas Drilling & Production Activities." www.shaleinfo.ohio.gov/Basics/SummaryofRegulatoryAuthority.aspx#.
Shauk, Zain. 2014. "U.S. Natural Gas Exports Will Fire Up in 2015." *Bloomberg Business*, November 6. www.bloomberg.com/bw/articles/2014-11-06/u-dot-s-dot-natural-gas-exports-will-fire-up-in-2015.
Sherman, Daniel. 2011. *Not Here, Not There, Not Anywhere*. Washington, DC: Resources for the Future Press.
Shingler, Dan. 2015. "Running on Empty: Some Say Utica Downturn Is All Part of the Cycle." *Crain's Business Cleveland*, July 25. www.crainscleveland.com/article/20150725/NEWS/150729870/running-on-empty-some-say-utica-downturn-is-all-part-of-the-cycle.
———. 2015a. "Shale is Suffering: Declining Oil and Natural Gas Prices Have Led to Plenty of Regional Cutbacks." *Crains Cleveland Business*, February 8. www.crainscleveland.com/article/20150208/SUB1/302089982/shale-is-suffering-declining-oil-and-natural-gas-prices-have-led-to.
Shipp, Brett. 2015. "Arlington Officials Report on Fracking Fluid Blowout." *WFAA*, June 17. www.wfaa.com/story/news/local/tarrant-county/2015/06/16/arlington-officials-report-on-fracking-fluid-blowout/28844657/.
Sider, Alison. 2014. "Demand for Sand Takes Off Thanks to Fracking." *Wall Street Journal*, August 4. www.wsj.com/articles/demand-for-sand-takes-off-thanks-to-fracking-1407193760.
Simmers, Richard. 2012. "2012 Ohio Oil and Gas Summary." http://oilandgas.ohiodnr.gov/portals/oilgas/pdf/O&GAnnRep_2012_FIN.pdf.
Sinnott, Molly. 2012. "Urban Oil and Gas Wells Pop Up across Colorado, but Number of Permits Decreasing." *Colorado Public News*, July 5. www.cpt12.org/news/index.php/urban-oil-and-gas-wells-pop-up-across-colorado/.
Sixel, L. M. 2015. "New Figures Suggest Area Could Lose Jobs for the Year." *Houston Chronicle*, September 18. www.houstonchronicle.com/business/economy/article/Houston-lost-energy-manufacturing-jobs-in-August-6513800.php.
Sjoberg, Lennart. 2000. "Factors in Risk Perception." *Risk Analysis* 20(1): 1–11.
Skolnick, David. 2015. "Fracking Opponents Are Willing to Put Money Where Mouths Are." *The Vindicator*, November 13. www.vindy.com/news/2015/nov/13/fracking-opponents-are-willing-to-put-mo/#sthash.LuTpP6yc.dpuf.
Slonecker, E. T., L. E. Milheim, C. M. Roig-Silva, A. R. Malizia, D. A. Marr, and G. B. Fisher. 2012. "Landscape Consequences of Natural Gas Extraction in Bradford and Washington Counties, Pennsylvania, 2004–2010." *U.S. Geological Survey Open-File Report 2012–1154*. http://pubs.usgs.gov/of/2012/1154/

Sloto, Ronald A. 2013. "Baseline Groundwater Quality from 20 Domestic Wells in Sullivan County, Pennsylvania, 2012." *USGS Sci. Investigations Report, 2013–5085*. http://pubs.usgs.gov/sir/2013/5085.
Slovic, Paul. 1987. "Perception of Risk." *Science* 235(4799): 280–285.
Smith, Carrie. 2014. "Fracking Waste Could be Headed Here." *Cincinnati Enquirer*, January 14. http://archive.cincinnati.com/article/20140114/NEWS01/301140008/Fracking-waste-could-headed-here
Smith, Terry. 2015. "Athens Has More Than One Law in Its Anti-Fracking Arsenal." *The Athens News*, May 6. www.athensnews.com/opinion/wearing_thin/athens-has-more-than-one-law-in-its-anti-fracking/article_22fd2555-ccad-58bd-9dbc-eabd0efd1055.html.
———. 2015a. "State Supreme Court Denies Driller's Request for Order Saying Local Zoning Can't Restrict Fracking: Court Ruling Doesn't Include Explanation, So Its Significance Is Uncertain." *The Athens News*, November 10. www.athensnews.com/news/local/state-supreme-court-denies-driller-s-request-for-order-saying/article_6100f6d6-87df-11e5–92cf-f37e50c060af.html.
———. 2015b. "U.S. EPA Praises Ohio's Injection-Well Program; Critics Not Impressed." *The Athens News*, September 20. www.athensnews.com/news/local/u-s-epa-praises-ohio-s-injection-well-program-critics/article_01e15e52-5fbc-11e5-a5e9-c3dccc76aa76.html.
Solomon, Dean, and Kurt H. Schindler. 2012. "Can Local Governments Regulate Oil and Gas Development? Michigan Communities Are Considering Their Roles in Regulating Oil and Gas Development through Zoning—Just What Michigan's Laws Permit Local Governments to Regulate is Complicated." Michigan State Extension, March 22. http://msue.anr.msu.edu/news/can_local_governments_regulate_oil_and_gas_development.
Song, Lisa, Jim Morris, and David Hasemyer. 2014. "Fracking Boom Leaves Texans under a Toxic Cloud." *Bloomberg News*, February 20. www.bloomberg.com/news/2014-02-20/fracking-boom-leaves-texans-under-a-toxic-cloud.html
Songer, Joe. 2014. "Can Natural Gas Bring Back U.S. Factory Jobs? Obama Touted the Energy Boom in His State of the Union as Fuel for Economic Revival." *National Geographic*, January 31. http://news.nationalgeographic.com/news/energy/2014/01/140131-natural-gas-manufacturing-jobs/.
Soraghan, Mike. 2015. "The Fracking 'Loophole' That Just Keeps Growing." *Environment and Energy Publishing*, August 18. www.eenews.net/stories/1060023558.
———. 2015a. "Wastewater Spills Untracked in Texas, Masking Pollution." *Environment and Energy Publishing*, November 18. www.eenews.net/energywire/2015/11/18/stories/1060028166.
———. 2014. "Spills up 17 Percent in U.S. in 2013." *Environment and Energy Publishing*, May 12. www.eenews.net/stories/1059999364.
———. 2012. "Texas Inspectors Use Their Strongest Punishment against Paperwork Violations." *Environment and Energy Publishing*, April 4. www.eenews.net/stories/1059962437.
———. 2011. "Puny Fines, Scant Enforcement Leave Drilling Violators with Little to Fear." *Environment and Energy Publishing*, November 14. www.eenews.net/stories/1059956366.
Southern Illinoisans Against Fracturing Our Environment [SAFE]. 2013. "Home." www.dontfractureillinois.net/

Spector, Harlan. 2012. "As Fracking Debate Heats Up, Broadview Heights Already Shows Strains of Oil, Gas Well Drilling." *Cleveland Plain-Dealer*, June 11. www.cleveland.com/metro/index.ssf/2012/06/as_fracking_debate_heats_up_br.html.

State Journal Staff. 2012. "Fracking Ban Overturned, Morgantown Considers Zoning Instead." *State Journal*, February 22. www.statejournal.com/story/16580946/fracking-ban-overturned-morgantown-considers-zoning-instead.

Stephan, Mark. 2002. "Environmental Information Disclosure Programs: They Work, But Why?" *Social Science Quarterly* 83(1): 190–205.

Stephens, Ross G. 1974. "State Centralization and the Erosion of Local Autonomy." *Journal of Politics* 36(1): 44–76.

Stever, James A. 2005. "Adapting Intergovernmental Management to the New Age of Terrorism." *Administration & Society* 37(4): 379–403.

Stoker, Gerry. 1998. "Governance as Theory: Five Propositions." *International Social Science Journal* 50(155): 17–27.

Stokols, Eli. 2015. "Oil and Gas Task Force Offers Few Solutions to Local Control Issue." *KDVR*, February 24. http://kdvr.com/2015/02/24/oil-and-gas-task-force-offers-few-solutions-to-local-control-issue/.

———. 2013. "Fort Collins Approves Fracking Ban, Defying Hickenlooper's Threat." *KDVR*, March 6. http://kdvr.com/2013/03/06/fort-collins-approves-fracking-ban-defying-hickenloopers-threat/.Storrow, Benjamin. 2015. "Wyoming Sues Interior Department over Rules for Drilling on Public Land." *Casper Star Tribune*, March 26. http://trib.com/business/energy/wyoming-sues-interior-department-over-rules-for-drilling-on-public/article_2393b613-b31b-5f21-9f17-fec1a37d5af9.html

Sun Bay Staff. 2015. "Rodrigues Swipes at Home Rule: Takes Aim at Municipal Fracking Bans." *Sun-Bay Paper*, September 25. www.sunbaypaper.com/florida/environment/rodrigues-swipes-at-home-rule-takes-aim-at-municipal-fracking-bans.

Susman, Tina. 2015. "New York's Fracking Ban Leads to Secession Talk." *Governing*, March 26. www.governing.com/topics/transportation-infrastructure/new-yorks-fracking-ban-lead-to-succession.html.

Sustainable Shale. 2016. "What We Do?" www.responsibleshaledevelopment.org/what-we-do/performance-standards/.

Svara, James H., Tanya C. Watt, and Hee Soun Jang. 2013. "How Are US Cities Doing Sustainability? Who Is Getting on the Sustainability Train, and Why?" *Cityscape* 15(1): 9–44.

Tenser, Phil. 2014. "Fracking Proposals Withdrawn from Contention for November Ballot." *The Denver Channel*, August 6. www.thedenverchannel.com/news/politics/fracking-proposals-withdrawn-from-contention-for-november-ballot.

Texas Alliance of Groundwater Districts. 2013. "More on EAA v. Bragg." www.texasgroundwater.org/news-events/item/46-more-on-eaa-v-bragg.html.

Texas Commission on Environmental Quality [TCEQ]. 2016. "Oil and Gas Facilities: Air Compliance Information." www.tceq.texas.gov/assistance/industry/oil-and-gas/oilgas_air.html.

———. 2015. "Who Regulates Oil and Gas Activities in Texas?" www.tceq.texas.gov/assets/public/assistance/sblga/oil-gas/statewide_oilgas_prog_info.pdf.

———. 2014. "History." www.tceq.texas.gov/about/tceqhistory.html (January 4, 2014).

———. 2014a. "Organization." www.tceq.texas.gov/about/organization/ed.html.
———. 2014b. "Texas Active Oil and Gas Wells—January 2014." www.tceq.state.tx.us/assets/public/implementation/barnett_shale/bs_images/txOilGasWells.png.
———. 2014c. "Water Rights." www.tceq.texas.gov/permitting/water_rights/wr_amiregulated.html.
Texas Comptroller of Public Accounts. 2014. "Sources of Revenue." www.texastransparency.org/State_Finance/Revenue/Sources/2015/Sources_of_Revenue_2015.pdf.
———. "Quarterly Sales Tax Historical Data." https://mycpa.cpa.state.tx.us/allocation/HistSales.jsp.
Texas House Committee on Business & Industry. 2015. "Economics and Fiscal Information on Oil and Gas Industry." www.legis.state.tx.us/tlodocs/84R/handouts/C0402015022411001/43bcd96b-8a58-4f5b-bb6d-8f5474b10cbc.PDF.
Texas Oil and Gas Association [TXOGA]. 2013. "Economic Impact." www.txoga.org/resources/economic-impact/.
———. 2013a. "Fueling the Texas Economy." www.txoga.org/assets/doc/PRINTABLE_2013_FUELING_THE_TEXAS_ECONOMY_download.pdf.
Texas Secretary of State. 2014. "Voter Registration and Unofficial Early Voting Figures." www.sos.state.tx.us/elections/historical/counties.shtml#A (March 14, 2014).The Post-Standard. 2011. "Marcellus Shale Links: Energy Group Says Oil and Gas Drilling Can Rescue Economy." *The Post-Standard*, March 29. http://www.syracuse.com/news/index.ssf/2011/03/marcellus_shale_links_energy_g.html
Tierney Kathleen J., Michael K. Lindell, and Ronald W. Perry. 2001. *Facing the Unexpected: Disaster Preparedness and Response in the United States*. Washington, DC: Joseph Henry Press.
Tita, Bob. 2015. "New Rules for Tank Cars Sent to White House for Review Proposed Regulations Come in Response to Series of Oil-Train Accidents." *Wall Street Journal*, February 5. www.wsj.com/articles/new-rules-for-tank-cars-sent-to-white-house-for-review-1423195189.
Tomain, Joseph P., and Richard D. Cudahy. 2004. *Energy Law in a Nutshell*. St. Paul, MN: West Publishing.
Town of DISH. 2014. "Odor Ordinance." http://townofdish.com/objects/Odor_Ordinance.doc.
———. 2014a. "Oil and Gas Ordinance." http://townofdish.com/objects/06-01A_OilGasOrdinance.pdf.
———. 2014b. "Ordinance 08–09a Gas Drilling Amended." http://townofdish.com/objects/ORD_08-09A_Gas_Drilling_Ord._Amended.doc.
Townsend-Small, Amy. N.d. "UC Groundwater Research of Ohio." www.artsci.uc.edu/departments/geology/research-and-scholarship/UCGWResearchofOhio.html.
Trisolini, Katherine A. 2010. "All Hands on Deck: Local Governments and the Potential for Bidirectional Climate Change Regulation." *Stanford Law Review* 62(March): 669–746.
Udell, Erin. 2013. "Fracking Brings Revenue, Concerns to Windsor Long-Term Health, Environment Effects a Concern for Town Board Member." *Windsor Beacon*, July 18. http://archive.coloradoan.com/article/20130719/WINDSORBEACON01/307190010/Fracking-brings-revenue-concerns-Windsor.

Unger, Todd. 2015. "Arlington Rejects New Drilling Lease for Lake Arlington." *WFAA*, November 11. www.wfaa.com/story/news/local/tarrant-county/2015/11/11/arlington-rejects-new-drilling-lease-for-lake-arlington/75553580/.

Union of Concerned Scientists. N.d. "Environmental Impacts of Natural Gas." www.ucsusa.org/clean_energy/our-energy-choices/coal-and-other-fossil-fuels/environmental-impacts-of-natural-gas.html#.VRrnAlzyN1I.

———. N.da. "Natural Gas Flaring, Processing, and Transportation." www.ucsusa.org/clean_energy/our-energy-choices/coal-and-other-fossil-fuels/natural-gas-flaring-processing-transportation.html#.VhbeGhNViko.

———. N.db. "The Future of Natural Gas." www.ucsusa.org/clean_energy/our-energy-choices/coal-and-other-fossil-fuels/the-future-of-natural-gas.html#.Vx-n5GOG_ZY.

United States Census Bureau. 2016. "Akron, OH Quick Facts." www.census.gov/quickfacts/table/PST045215/3901000.

———. 2016a. "Arlington, TX Quick Facts." www.census.gov/quickfacts/table/PST045215/4804000.

———. 2016b. "Broadview Heights, OH Quick Facts." www.census.gov/quickfacts/table/PST045215/3909064.

———. 2016c. "Denton, TX Quick Facts." www.census.gov/quickfacts/table/PST045215/4819972.

———. 2016d. "Fort Collins, CO Quick Facts." http://quickfacts.census.gov/qfd/states/08/0827425.html.

———. 2016e. "Fort Worth, TX Quick Facts." www.census.gov/quickfacts/table/PST045215/4827000.

———. 2016f. "Greeley, CO Quick Facts." http://quickfacts.census.gov/qfd/states/08/0832155.html.

———. 2016g. "Longmont, CO Quick Facts." http://quickfacts.census.gov/qfd/states/08/0845970.html.

———. 2016h. "Loveland, CO Quick Facts." http://quickfacts.census.gov/qfd/states/08/0846465.html.

———. 2016i. "Mansfield, TX Quick Facts." www.census.gov/quickfacts/table/PST045215/4846452.

———. 2016j. "Munroe Falls, OH Quick Facts." www.census.gov/quickfacts/table/PST045215/3953312.

———. 2016k. "Youngstown, OH Quick Facts." www.census.gov/quickfacts/table/PST045215/3988000.

———. 2015. "Fact Finder." http://factfinder.census.gov/faces/tableservices/jsf/pages/productview.xhtml?src=bkmk.

———. 2014. "State Government Tax Collections." www.census.gov/govs/statetax/.

———. 2013. "2013 Annual Survey of State Government Tax Collections." www2.census.gov/govs/statetax/2013stcreport.pdf.

United States Congress. 2015. "Senate Bill 828." www.congress.gov/bill/114th-congress/senate-bill/828.

———. 2014. "H.R.4155—Fight Russian Energy Exploitation (FREE) Act." www.congress.gov/bill/113th-congress/house-bill/4155.

———. 2014a. "Summary: H.R.6—Domestic Prosperity and Global Freedom Act." www.congress.gov/bill/113th-congress/house-bill/6?q=%7B%22search

%22%3A%5B%22Domestic+Prosperity+and+Global+Freedom+Act%22%5D%7D&resultIndex=2.

United States Department of Transportation [DOT]. 2015. "Briefing." www.transportation.gov/briefing-room/final-rule-on-safe-rail-transport-of-flammable-liquids.

———. 2015a. "Rule Will Make Significant and Extensive Changes to Improve Accident Prevention, Mitigation, and Emergency Response." www.transportation.gov/briefing-room/final-rule-on-safe-rail-transport-of-flammable-liquids#sthash.HRNor8nU.dpufwww.transportation.gov/briefing-room/final-rule-on-safe-rail-transport-of-flammable-liquids.

United States Forest Service [USFS]. 2015. "Administration of Oil and Natural Gas Activities: Wayne National Forest System Lands—Administration of Oil and Natural Gas Activities." www.fs.usda.gov/detail/wayne/home/?cid=stelprdb5376502.

United States Geological Survey [USGS]. 2016. "Induced Earthquakes." http://earthquake.usgs.gov/research/induced/.

———. 2014. "Earthquake Facts & Earthquake Fantasy." http://earthquake.usgs.gov/learn/topics/megaqk_facts_fantasy.php.

United States Treasury. 2016. "United States – Progress Report on Fossil Fuel Subsidies." www.treasury.gov/open/Documents/USA%20FFSR%20progress%20report%20to%20G20%202014%20Final.pdf.

University of Texas Libraries. 2014. "Railroad Commission of Texas, Oil and Gas Division: An Inventory of Oil and Gas Division Correspondence and Reports at the Texas State Archives, 1890–1943, Bulk 1919–1935." www.lib.utexas.edu/taro/tslac/10227/tsl-10227.html.

U.S. Census Bureau. N.d. "Selected Historical Decennial Census Population and Housing Counts." www.census.gov/population/www/censusdata/hiscendata.html.

Vardon, Joe. 2013. "New Severance-Tax Plan Offers Aid to 33 Counties." *The Columbus-Dispatch*, June 7. www.dispatch.com/content/stories/local/2013/06/06/Kasich-reviving-frack-tax-plan.html (January 19, 2014).

Veksin, Alison. 2013. "California Fracking Fight Has $25 Billion Taxes at Stake." *Bloomberg News*, March 17. www.bloomberg.com/news/articles/2013-03-17/california-fracking-fight-has-25-billion-taxes-at-stake.

Vengosh, Avner, Robert B. Jackson, Nathaniel Warner, Thomas H. Darrah, and Andrew Kondash. 2014. "A Critical Review of the Risks to Water Resources from Unconventional Shale Gas Development and Hydraulic Fracturing in the United States." *Environmental Science and Technology* 48(15): 8334–8348.

Vig, Norman, and Michael E. Kraft. 2009. "Introduction." In *Environmental Politics & Policy*. 7th ed. eds. Norman Vig and Michael E. Kraft. Washington, DC: CQ Press.

Wagner, John. 2014. "O'Malley Says He Is Ready to Allow 'Fracking' in Western Maryland, with Strict Safeguards." *Washington Post*, November 25. www.washingtonpost.com/local/md-politics/omalley-says-he-is-ready-to-allow-fracking-in-western-maryland-with-strict-safeguards/2014/11/25/36234f34-74b9-11e4-9d9b-86d397daad27_story.html.

———, and Scott Clement. 2015. "Poll: Marylanders Like Hogan, but Oppose Effort to Slow School Funding." *Washington Post*, February 10. www.washingtonpost.com/local/md-politics/poll-marylanders-like-hogan-but-oppose-effort-to-slow-school-funding/2015/02/10/d834928a-b099-11e4-827f-93f454140e2b_story.html.

Walker, David B. 1995. *The Rebirth of Federalism: Slouching toward Washington*. Chatham, NJ: Chatham House.

Wang, Robert. 2011. "Gov. Kasich says 'Fracking' Is Opportunity for Ohio." *The Massillion Independent*, February 11. www.indeonline.com/article/20110211/News/302119909.

Warner, Barbara, and Jennifer Shapiro. 2013. "Fractured, Fragmented Federalism: A Study in Fracking Regulatory Policy." *Publius: The Journal of Federalism* 43(3): 474–496.

Warner, Nathaniel R., Cidney A. Christie, Robert B. Jackson, and Avner Vengosh. 2013. "Impacts of Shale Gas Wastewater Disposal on Water Quality in Western Pennsylvania." *Environmental Science & Technology* 47(20): 11849–11857.

Warren, Bob. 2015. "St. Tammany Fracking Suit Moves to Baton Rouge Courtroom Monday." *New Orleans Times-Picayune*, April 20. www.nola.com/crime/index.ssf/2015/04/st_tammany_fracking_suit_moves.html.

Warsmith, Stephanie. 2015. "Akron Passes Resolution Opposing State Fracking Bill." *Akron Beacon-Journal*, July 13. www.ohio.com/news/break-news/akron-passes-resolution-opposing-state-fracking-bill-1.607900.

Weaver, Bridgett. 2015. "COGCC Has a Plan for Future Earthquakes." *Greeley Tribune*, February 10. www.greeleytribune.com/news/14763491-113/cogcc-has-a-plan-for-future-earthquakes.

Weber, Harry R. 2013. "British Utility to Buy LNG from Cheniere." *Houston Chronicle*, March 25. www.chron.com/business/energy/article/British-utility-to-buy-LNG-from-Cheniere-4383624.php.

Weber, Jeremy G., James Burnett, and Irene M. Xiarchos. 2016. "Broadening Benefits from Natural Resource Extraction: Housing Values and Taxation of Natural Gas Wells as Property." *Journal of Policy Analysis and Management* 35(3): 587–614.

Weinheimer, Dan. 2013. "Oil and Gas Questions and Answers from Citizens: Updated May 7, 2013." www.fcgov.com/common/pdfs/spotlight-pdf.php?id=527.

Welch, Terrence S. 2007. "Containing Urban Sprawl: Is Reinvigoration of Home Rule the Answer." *Vermont Journal of Environmental Law* 9: 131–153.

Wernau, Julie. 2015. "Illinois Misses the Fracking Boom because of Falling Oil Prices." *Chicago Tribune*, January 24. www.chicagotribune.com/business/ct-fracking-illinois-0125-biz-20150123-story.html#page=1.

———. 2013. "Illinois Counties Can Do Little but Brace for Fracking's Impact: Protections for Landowners in Proposed Law Only Go So Far." *Chicago Tribune*, March 17. http://articles.chicagotribune.com/2013-03-17/business/ct-biz-0317-fracking-illinois-20130317_1_oil-boom-illinois-counties-oil-and-gas.

Wertz, Joe. 2015. "Stillwater Approves New Oil and Gas Rules, Industry Says They Might Violate New Law." *National Public Radio*, July 21. https://stateimpact.npr.org/oklahoma/2015/07/21/stillwater-approves-new-oil-and-gas-rules-industry-says-they-might-violate-new-law/.

Western Natural Resources Law Group. 2014. "Severance Tax Research Project." www.leg.state.co.us/CLICS/CLICS2014A/commsumm.nsf/b4a3962433b52fa787256e5f00670a71/6e505a47c1f2abad87257cc3004f788e/$FILE/140423%20AttachC.pdf.

WFMJ. N.d. "Brookfield Manufacturer Lays Off 10% of the Workforce." www.wfmj.com/story/28022961/brookfield-company-lays-off-10-of-the-workforce.

Williams, Wayne. 2016. "Energy Measures Fail to Make November Ballot: Backers of No. 75 and 78 Fall Short." https://content.govdelivery.com/accounts/COSOS/bulletins/160232c.
Wilonsky, Robert. 2014. "Update: Mayor Rawlings Says of Trinity East's $30 Million Lawsuit, 'My Prediction Has Come True'." *Dallas Morning News*, February 13. http://cityhallblog.dallasnews.com/2014/02/denied-right-to-drill-in-dallas-trinity-east-energy-sues-city-to-recoup-its-19-million.html/.
Wiseman, Hannah. 2015. "Governing Fracking from the Ground Up." *University of Texas Law Review* 93: 29–46. www.texaslrev.com/wp-content/uploads/Wiseman-93-SeeAlso.pdf.
———. 2009. "Untested Waters: The Rise of Hydraulic Fracturing in Oil and Gas Production and the Need to Revisit Regulation." *Fordham Environmental Law Review* 20: 115.
Wittmeyer, Hannah. 2013. "Fracking Regulations in Texas." http://frackwire.com/texas-regulations/.
Wobbekind, Richard, Brian Lewandowski, and Ernie Chung. 2014. "Oil and Gas Industry Economic and Fiscal Contributions in Colorado by County, 2008–2012." www.api.org/~/media/files/news/2014/14-october/colorado-api-economic-impact-study.PDF.
Woock, Randy. 2010. "Colorado Groups Disagree on Importing N.Y. Fracking Moratorium." *Trinidad Times*, December 23. http://trinidad-times.com/colorado-groups-disagree-on-importing-ny-fracking-moratorium-p1323-1.htm.
Wood, Curtis. 2011. "Exploring the Determinants of the Empowered US Municipality." *State and Local Government Review* 43(2): 123–139.
Woods, Neal D., and Matthew Potoski. 2010. "Environmental Federalism Revisited: Second-Order Devolution in Air Quality Regulation." *Review of Policy Research* 27(6): 721–739.
Wright, Deil S. 1978. *Understanding Intergovernmental Relations: Public Policy and Participants' Perspectives on Local, State, and National Governments*. North Scituate, MA: Duxbury.
Wronski, Richard. 2014. "Trains Filled with Flammable Crude Oil Threaten Area, Records Show." *Chicago Tribune*, July 14. www.chicagotribune.com/news/ct-oil-trains-met-20140714-story.html.
Yachnin, Jennifer. 2016. "Pa. Dems Squabble over Fracking—and Guess Who's for It?" *Environment and Energy Publishing*, February 12. www.eenews.net/stories/1060032312.
Yeagle, Patrick. 2013. "What to Expect with Fracking: State Law Opens Up Southern Illinois to Hydraulic Fracturing." *Illinois Times*, June 27. http://illinoistimes.com/article-12566-what-to-expect-with-fracking.html.
Yi, Hongtao. 2013. "Clean Energy Policies and Green Jobs: An Evaluation of Green Jobs in US Metropolitan Areas." *Energy Policy* 56(May): 644–652.
Zarsky, Lyuba. 2010. "Climate-Resilient Industrial Development Paths: Design Principles and Alternative Models." In *Towards New Developmentalism: Market as Means Rather than Master*. eds. Shahrukh Rafi Khan and Jens Christiansen. Milton Park, Abingdon, United Kingdom: Routledge Economics.
Zielinska, Barbara, Dave Campbell, and Vera Samburova. 2014. "Impact of Emissions from Natural Gas Production Facilities on Ambient Air Quality in

the Barnett Shale Area: A Pilot Study." *Journal of the Air & Waste Management Association* 64(12): 1369–1383.
Zimmerman, Joseph F. 2012. *State-Local Governmental Interactions.* Albany, NY: SUNY Press.
———. 1995. *State-Local Relations: A Partnership Approach.* Westport, CT: Praeger Publishers.
Zremski, Jeremy. 2014. "Youngstown, Ohio, Is a City Changed by Fracking: Controversial as Fracking Is, Youngstown, Ohio, Embraces It as an Economic Savior—Eager to Take Risks That Have Even Caused the Earth to Tremble." *The Buffalo-News*, May 18. www.buffalonews.com/city-region/youngstown-ohio-is-a-city-changed-by-fracking-20140518.

All online resources were checked and utilized between 2013 and 2017.

Glossary

The politics of shale gas and oil sometimes involves technical jargon. Because of this, a variety of fracking-related terms and a brief explanation of their roles are offered as follows:

Blender A blender is a piece of equipment used by operators to mix the gels, fluids, and compounds needed to stimulate and fracture the shale rocks. The specific recipe is a function of the needs of the operator, the oil or gas deposit, and the 'specifics' of the geological formation.

Casing During well construction, operators install a pipe that is encased in cement while the well (and wellbore) is being built. The casing pipe prevents cross-contamination between fracking fluids and underground formations and is designed to prevent the well from caving in. The casing also helps operators control the fluids and internal pressures.

Flaring Flaring is performed when operators burn off excess natural gas, often to relieve high pressures within the well.

Flowback Water used during fracking operations that returns to the surface is called flowback.

Horizontal drilling Part of the production process by which the well turns horizontal, typically to follow an energy/mineral deposit.

Hydraulic fracturing (fracking) Part of the production process that involve the high-volume injection of water, sand, and proprietary chemicals.

Mineral estate The mineral estate includes the ownership and property rights associated with the subsurface resource.

Oilfield servicing companies Companies and firms that provide support services during production but do not own the mineral.

Produced waters Once the fracking procedure is completed, the pressure is reversed. Water that emerges from the formation along with oil and gas is considered produced waters.

Proppant Industry uses proppant, typically sand, to keep the fissures or cracks open once the fracking fluid is withdrawn. This enables the oil and gas to flow from the formation and make its way to the surface.

Reverse setback Mandated by the applicable regulatory agency, a reverse setback is the required distance or buffer space required between ***new*** commercial and residential development and existing wells.

Rig (drilling) Operators use drilling rigs as part of the well construction process. A drilling rig is the machine that drills the wellbore (i.e., the hole used to reach the shale formation). The rig includes mud tanks, mud pumps, derricks, and other power-generating machinery needed to manage the well pad.

Separation unit Once the oil/gas flows to the surface, it must be separated from other liquids. Operators use separation units to separate the hydrocarbons.

Setback Imposed by the applicable regulatory agency, a setback is the required distance or buffer space between a ***new*** well and existing commercial and residential development.

Shale Operators drill into shale, which is a type of sedimentary rock. Oil and gas deposits may settle into formations (about 1 percent of the actual rock formation), which typically forms fairly impermeable but thin layers.

Surface estate The surface estate is ownership and attendant rights of the surface resources.

Tank, tank battery, or battery Industry uses tanks or a tank battery, which is a group of storage containers to store produced liquids.

Tight oil Oil that is produced from sandstones, limestones, and shale formations.

Wellbore The wellbore is the drill hole or borehole made by the drilling rig. It also includes the uncased portion of the drilled well.

Well completion Once a well is completed, it is ready to start producing oil or gas.

Well pads Industry drills on parcels of land are called well pads. Pads can include one or more surface leases and can have multiple wells/wellbores located on them.

Wellhead A wellhead includes the surface-level aspects of the well such as valves and seals. It is designed to prevent leaks and spills as oil and gas flow upward.

Wildcatters A term used to describe operators who drill exploratory wells.

Sources: Dunn (2014), Encana (2013), EPA (2011)

Index

For Product Safety Concerns and Information please contact our EU
representative GPSR@taylorandfrancis.com
Taylor & Francis Verlag GmbH, Kaufingerstraße 24, 80331 München, Germany

www.ingramcontent.com/pod-product-compliance
Ingram Content Group UK Ltd.
Pitfield, Milton Keynes, MK11 3LW, UK
UKHW020133200926
475702UK00005B/728

* 9 7 8 1 4 9 8 7 4 2 4 1 2 *